T0282633

CAMBRIDGE LIBRARY COLLECTION

Books of enduring scholarly value

Botany and Horticulture

Until the nineteenth century, the investigation of natural phenomena, plants and animals was considered either the preserve of elite scholars or a pastime for the leisured upper classes. As increasing academic rigour and systematisation was brought to the study of 'natural history', its subdisciplines were adopted into university curricula, and learned societies (such as the Royal Horticultural Society, founded in 1804) were established to support research in these areas. A related development was strong enthusiasm for exotic garden plants, which resulted in plant collecting expeditions to every corner of the globe, sometimes with tragic consequences. This series includes accounts of some of those expeditions, detailed reference works on the flora of different regions, and practical advice for amateur and professional gardeners.

Flowers of the Field

A keen collector and sketcher of plant specimens from an early age, the author, educator and clergyman Charles Alexander Johns (1811–74) gained recognition for his popular books on British plants, trees, birds and countryside walks. *The Forest Trees of Britain* (1847–9), one of several works originally published by the Society for Promoting Christian Knowledge, is also reissued in this series. First published by the Society in 1851, Johns' best-known work is this two-volume botanical guide to common British flowering plants. Following the Linnaean system of classification, Johns describes the various plant families, providing the common and Latin names for each species. The work is especially noteworthy for its delicate and meticulous line drawings, based on watercolours by the botanical artist Emily Stackhouse and the author's sisters Julia and Emily. Volume 1 includes an introduction to the Linnaean system and the botanical terms used in the text.

Cambridge University Press has long been a pioneer in the reissuing of out-of-print titles from its own backlist, producing digital reprints of books that are still sought after by scholars and students but could not be reprinted economically using traditional technology. The Cambridge Library Collection extends this activity to a wider range of books which are still of importance to researchers and professionals, either for the source material they contain, or as landmarks in the history of their academic discipline.

Drawing from the world-renowned collections in the Cambridge University Library and other partner libraries, and guided by the advice of experts in each subject area, Cambridge University Press is using state-of-the-art scanning machines in its own Printing House to capture the content of each book selected for inclusion. The files are processed to give a consistently clear, crisp image, and the books finished to the high quality standard for which the Press is recognised around the world. The latest print-on-demand technology ensures that the books will remain available indefinitely, and that orders for single or multiple copies can quickly be supplied.

The Cambridge Library Collection brings back to life books of enduring scholarly value (including out-of-copyright works originally issued by other publishers) across a wide range of disciplines in the humanities and social sciences and in science and technology.

Flowers of the Field

VOLUME 1

CHARLES ALEXANDER JOHNS

CAMBRIDGE
UNIVERSITY PRESS

University Printing House, Cambridge, CB2 8BS, United Kingdom

Published in the United States of America by Cambridge University Press, New York

Cambridge University Press is part of the University of Cambridge.
It furthers the University's mission by disseminating knowledge in the pursuit of
education, learning and research at the highest international levels of excellence.

www.cambridge.org
Information on this title: www.cambridge.org/9781108068642

© in this compilation Cambridge University Press 2014

This edition first published 1853
This digitally printed version 2014

ISBN 978-1-108-06864-2 Paperback

FLOWERS OF THE FIELD.

BY THE

REV. C. A. JOHNS, B.A. F.L.S.

AUTHOR OF BOTANICAL RAMBLES, THE FRUIT TREES OF BRITAIN,
ETC. ETC.

IN TWO VOLUMES.

VOL. I.

PUBLISHED UNDER THE DIRECTION OF
THE COMMITTEE OF GENERAL LITERATURE AND EDUCATION,
APPOINTED BY THE SOCIETY FOR PROMOTING
CHRISTIAN KNOWLEDGE.

LONDON:
PRINTED FOR THE
SOCIETY FOR PROMOTING CHRISTIAN KNOWLEDGE;
SOLD AT THE DEPOSITORY,
GREAT QUEEN STREET, LINCOLN'S INN FIELDS,
4, ROYAL EXCHANGE, 16, HANOVER STREET, HANOVER SQUARE;
AND BY ALL BOOKSELLERS.

LONDON:
R. CLAY, PRINTER, BREAD STREET HILL.

INTRODUCTION

TO

BRITISH BOTANY.

———◆———

CHAPTER I.

EXPLANATION OF TERMS.

THOUGH the highest claim of these volumes is to intro-
duce the lover of Nature to an acquaintance with the
common British plants, the Author has given to his
first Chapter the somewhat presuming title of an "*Introduction to British Botany;*" lest those into whose hands
the work may fall should pass over the earlier part of
it as a treatise or summary of contents unconnected
with what follows, the perusal of which may, therefore, be omitted or deferred with safety. So far is
this from being the case, that the reader who is unac-
quainted with the elements of botany will find the body
of the work of little use, unless he carefully peruses the
earlier pages, and makes himself thoroughly acquainted
with the general plan.

The limits of a work of this kind will not allow any
account of the internal structure of plants, or of the
functions of their various organs. Nor, indeed, is such
description necessary in a work which professes merely
to teach the unscientific how to find out the names of
the flowers they may happen to fall in with in the
course of their country rambles. Such a knowledge of
plants as this, it may be said, and said with truth, is
not Botany ; nevertheless, it is a step towards Botany :
for there can be no doubt that scientific treatises on this
subject would often be studied with pleasure, if the

reader were familiar with simply the outward appearance
of the examples quoted : just as we take greater interest
in accounts of astronomical discoveries, if we have seen
and handled a telescope, than if we had merely had one
described to us, no matter with what accuracy and
minuteness. The reader, then, or, inasmuch as even the
elementary knowledge of a science can only be attained
by study, the *student* who wishes to make these volumes
practically useful in enabling him to find out the names
of our common wild flowers, is recommended to read
with care and attention the following pages, into which
the author has introduced nothing but what is essential
to the proper understanding of the body of the work, and
so to the attainment of his object.

Before a novice can commence the study of any science,
he must make himself acquainted with the terms employed by writers on that science ; he must not be
frightened if things new to him should have strange
names. Unmeaning and hard to be remembered they must
appear to him at first, but this will be only as long as
they remain mere sounds. When he has gained a knowledge of the *things* for which they stand, they will lose
their formidable appearance, and, hard as they may still
be to pronounce, they will very soon become familiar to
the mind, if not to the tongue. In a scientific treatise
on Botany, taken in its widest sense, these terms must of
necessity be very numerous. Not so, however, with a
popular description of the plants growing wild in a single
country of limited extent ; the Author, therefore, has
endeavoured to keep technical terms out of sight, as
much as possible, in the hope that the lover of Nature
may be beguiled into forming an acquaintance with the
outward appearance of the plants of his neighbourhood,
and eventually be induced to study their characters, or
to extend his researches beyond the limits of his own
country. He has, consequently, avoided the use of Latin
words wherever English ones would do as well, and has
often preferred to express by several words what might

have been defined by one, because that one was probably
strange to the reader. With respect to the organs of
plants, he has not noticed the existence of any but those
with which it is necessary that the student should be
familiar before he refers to the body of the work for a
description of any plant which he may have found;
these, with their principal peculiarities, may be described
at once. They are, ROOT, STEM, LEAF, STIPULES, BRACTS,
FLOWER, CALYX, COROLLA, STAMENS, PISTILS, FRUIT,
SEED, RECEPTACLE, and NECTARY.

THE ROOT.—The most frequent form of the *root* is a
tuft of fibres, each of which ends in a porous substance
serving to absorb moisture from the soil. In many
instances, however, the nourishment thus obtained,
instead of being transmitted at once to that part of the
plant which rises above the ground, is lodged in another
organ, which, though partaking in some measure the
properties of root and stem, is distinct from both. This,
too, with the fibres attached to it, is called a *root*, the
fibres themselves being named *rootlets*. The principal
forms of the root are :—

The *Creeping Root*, familiar examples of which are
afforded by Couch-grass and Horse-radish.

The *Spindle-shaped Root ;* examples, Carrot and
Parsnep.

A Spindle-shaped Root which ends abruptly, is termed
premorse (bitten off), as in Premorse Scabious, vol. i.
p. 315.

The *Tuberous Root* consists of one or more roundish,
solid masses, having the power of producing rootlets and
buds from several parts of its surface, as the Potato.

The *Bulbous Root* is a solid roundish mass, producing
rootlets at the lower extremity, and a bud at the other ;
it consists either of fleshy scales, as in the White Lily,
concentric circles, as in the Onion, or is of one uniform
substance throughout, as in the Crocus. This last is
sometimes called a *corm*.

THE STEM.—The stem is said to be *simple* when it

bears leaves, or leaves and flowers only without branches;
as in Grass of Parnassus, vol. i. p. 120.

A *compound stem* is repeatedly and irregularly
branched, as in Flax-seed, vol. i. p. 108.

A stem is said to be forked when it divides into two
branches of equal, or nearly equal, size. The stem of
Annual Knawel, vol. i. p. 229, is repeatedly forked.

The term *erect*, when applied to the stem, has the
same meaning as *perpendicular*.

An *ascending stem* is one which is horizontal when
first it leaves the root, and then becomes erect. When
several stems grow from one root, the central one is often
erect, the rest *ascending*, as in the common Mallow.

A *prostrate stem* trails along the ground without ever
becoming erect.

A *creeping stem* differs from the last by sending out
roots from its joints. Some plants have erect stems
with creeping *scions*, or shoots from the base, as the
Creeping Buttercup, vol. i. p. 11.

THE AXIL.—This name is given to the angle formed
by a leaf where it leaves the stem. A bud or flower
which springs from this angle, is termed *axillary*.

THE LEAF.—*Leaves* which spring directly from the
root are called *radical;* those which grow on the stem
are either *alternate*, as in Balsam, vol. i. p. 130; *opposite*, as in the Pink, p. 88; or *whorled:* the leaves of
Woodruff, p. 306, grow in *whorls*.

Leaves which have no stalks are termed *sessile* (sitting), as in Eryngo, vol. i. p. 254.

A leaf which consists but of one piece is said to be *simple*, as in Marsh Marigold, vol. i. p. 15 ; a *ternate leaf* consists of three *leaflets* on a common stalk, as in Medick, p. 149 ; a *quinate*, of five, as in Marsh Cinquefoil, p. 189. Other forms of the *compound leaf* are the pinnate, (from *penna*, a feather,) when a number of leaflets are ranged along the opposite sides of a common stalk, as in Saint-foin, vol. i. p. 169.

A simple leaf is sometimes *wavy* at the edge, as in the Oak, vol. ii. p. 185; 3-, 5-, or 7-*lobed*, as in the Mallows, pp. 110—113 ; and these *lobes* are often deeply *cut*, as in Geranium, p. 125. A leaf of five or more narrow lobes united near the main stalk, is termed *palmate*, (from *palma*, the palm of the hand,) as in Hellebore, p. 17. The *pedate* leaf differs from the palmate, in having the two side lobes divided a second time at the edge nearest the stalk. A leaf which is lobed after the manner of a pinnate leaf, is termed *pinnatifid*, (from *penna*, a feather, and *findo*, to cleave.)

If a stalk is attached to a leaf at or near its centre, such a leaf is termed *peltate*, (from *pelta*, a buckler,) as in Cotylédon, vol. i. p. 231.

A leaf through which a stalk passes is termed *perfoliate*, (from *per*, through, and *folium*, a leaf,) as in Hare's-ear, vol. i. p. 267.

Two leaves united by their bases, and allowing the stem to pass through them, are termed *connate*, (from *con*, together, and *nascor*, to grow,) as in Chlora, vol. ii. p. 37.

The margin of the leaf is either *entire*, as in Soapwort, vol. i. p. 90 ; *crenate*, as in Marsh Pennywort, p. 252 ; *serrate* (saw-edged), as in Rose, p. 199 ; *toothed*, as in Enchanter's Nightshade, p. 211 ; or *fringed*, as in Rockrose, p. 73.

With respect to form, the varieties of leaves are very numerous, and the terms employed to define them not less so. Those which occur in this volume are :—

Hair-like, or *capillary*, as in Fennel, vol. i. p. 271.

Linear, as in the Grasses and Pink, vol. i. p. 88.

Strap-shaped, as in Corrigiola, vol. i. p. 227.

Oblong, as in Rock-rose, vol. i. p. 73.

Elliptical, oval, with both ends alike, as in the leaflets of Rose, vol. i. p. 199.

Egg-shaped, oval, with the base broader than the extremity, as in Pear, vol. i. p. 200.

Inversely egg-shaped, oval, with the base narrower than the extremity, as in Brookweed, vol. ii. p. 87.

Rounded, as in Pyrola, vol. ii. p. 21.

Heart-shaped, as in Violet, vol. i. p. 76.

Inversely heart-shaped, as in the leaflets of Medick, vol. i. p. 149.

Kidney-shaped, as in Ground Ivy, vol. ii. p. 114.

Arrow-shaped, as in Tower Mustard, vol. i. p. 58.

Halbert-shaped, arrow-shaped, but with the barbs turned outwards.

Angular, as in Danish Scurvy-Grass.

DANISH SCURVY-GRASS.

Sword-shaped, as in Iris, vol. ii. p. 206.

STIPULES.—The base of the leaf-stalk is not unfrequently furnished with two sheathing wings; these are called stipules. The leaf of the Rose has oblong stipules at its base.

BRACTS.—Beneath the flower are frequently situated small leaves called *bracts.* Sometimes they are mere scales, as in Broom-rape, vol. ii. p. 68; but more frequently they are only to be distinguished from true leaves by their smaller size, as in Evening Primrose, vol. i. p. 210.

In the Umbelliferous Tribe, vol. i. p. 244, they often grow, several in a whorl, at the base of the general and partial umbels; and in Compound Flowers, vol. i. p. 317, they are yet more numerous at the base of the heads of flowers. When they grow in this form, they are termed an *involucre,* (from *involvo,* to wrap up, because they enclose the flowers before expansion.)

THE FLOWER.—This, as it is the most ornamental, so it is the most important part of the plant, being rarely produced until the juices fit for its nourishment have been selected by the roots and matured by the leaves, and containing all the apparatus necessary for perfecting seeds. In flowering plants, besides the parts which are indispensable to the ripening of seeds, there are others which evidently serve as a protection, and others, again, the use of which is not known. The flower, however, generally, being essential to the continuance of the species, has been selected, as the part on which to found every arrangement of plants which can lay claim to accuracy or utility. A thorough knowledge of its structure is therefore necessary before the student can proceed to discover the names of the commonest plants which are flung with so bountiful a hand over our hills and fields.

THE CALYX.—This name is given to that part of the flower which in the bud stage is outside all the rest, and which when the flower is expanded encircles the more delicate parts. It is usually green, and consists of several leaves, termed *sepals ;* but these sepals are often united at the base and form a cup, (hence the name *calyx*, a cup.)

It is unnecessary here to describe the various forms of the calyx, which are very numerous. It may be remarked, however, that when the calyx is divided into two distinct lobes, one of which overhangs the other, it is termed *gaping ;* in the Mallow Tribe, vol. i. p. 109, it is double ; and in Compound Flowers, p. 317, the Valerian, p. 308, and Teazel Tribes, p. 312, it is at first a mere ring, but afterwards becomes a chaffy or feathery appendage to the seed, termed a *pappus.*

THE COROLLA.—Within the calyx is the *corolla* (little crown), a ring of delicate leaves called *petals*, usually

coloured—that is, not green—and often fragrant. The petals are either distinct, as in the Rose, in which the

expanded part is termed the *limb*, the lower the *claw* ;
or united below, when the expanded part is termed the
border, the lower the *tube.* The corolla usually has as

many petals or divisions as there are sepals ; and if
these are all of the same size and shape, the corolla is
said to be *regular.*

The most common forms of the regular corolla of one
petal, are :—

Salver-shaped, as in Primrose, vol. ii. p. 129.

Funnel-shaped, as in Cowslip, vol. ii. p. 131.

Wheel-shaped, when the tube is no longer in proportion than the axle of a wheel, as in Speedwell, vol. ii.
p. 87.

Bell-shaped, as in Campanula, vol. ii. p. 3.

Trumpet-shaped, as in Convolvulus, vol. ii. p. 42.

When the irregular corolla of one petal is divided
into two lobes, one of which overhangs the other, it is
termed *labiate*, or *lipped*, as in the Natural Family
Labiatæ, vol. ii. p. 91 ; if the lips are open, it is said to
be *gaping*, as in Yellow Dead Nettle, vol. ii. p. 106 ; if
closed, *personate*, (from *persóna*, a mask,) as in Snap-
dragon, vol. ii. p. 74. In the Compound Flowers, vol. i.
p. 317, there are frequently two kinds of florets in one
flower, those of the *disk*, or centre, being tubular, without an evident border ; those of the *ray*, or margin,
strap-shaped, as in the Daisy.

Among *regular* flowers of many petals, the only form

which it will be necessary to mention here is the *cruci-form*, consisting of four petals placed cross-wise, as in the Cruciferous Tribe, vol. i. p. 35.

The most remarkable among the *irregular*, is the papilionaceous, (from *papilio*, a butterfly,) consisting of five petals, of which the upper one, called the *standard*, is usually the largest ; the two side ones are termed *wings*, and the two lower ones, which are often combined, form the *keel*, vol. i. p. 138.

Both calyx and corolla are not always found in the same flower, and when one only is present, it is sometimes difficult to decide by what name it should be called. In this case the term *perianth* (from the Greek *peri*, around, and *anthos*, a flower,) is a convenient one. Some flowers have neither calyx nor corolla, as Water Star-wort, vol. i. p. 215. When the *perianth* is said to be double, it is to be understood that calyx and corolla are both present.

THE STAMENS.—Within the perianth, and frequently attached to it, is a row of delicate organs called *stamens*, of which the lower part is termed the *filament*, the upper the *anther*. When the filament is slender throughout, it is said to be *thread-like ;* but if it be thick at the base, and taper to a point, it is said to be *awl-shaped*. The anther varies in shape, but is most frequently oblong, and composed of two lobes and as many cells, which are filled with a fine dust, called *pollen*. If there be no filament, the anther is said to be *sessile*. In a majority of flowers the number of stamens equals that of the petals; a few plants have but one stamen: very often the number of stamens is some multiple of the petals, that is, there are twice or thrice, &c. as many, and not a few flowers have from twenty to several hundred. Sometimes the filaments are united at the base into one or more sets, as in Hypericum, vol. i. p. 116 ; sometimes they form a hollow tube, the anthers being distinct, or *free*, as in Mallow, vol. i. p. 109 ; and sometimes the filaments are free, and the anthers are united into a ring, as in the Compound Flowers, vol. i. p. 317 ; and Heath, vol. ii. p. 14.

THE PISTIL.—This is the central part of the flower, and in its commonest form is a delicate column composed of three parts, the *ovary*, the *style*, and the *stigma*.

The *ovary*, (from *ovum*, an egg,) sometimes called the *germen*, contains the rudiments of the future seed.

The *style*, (from *stylos*, a column,) is to the pistil what the shaft is to a pillar, connecting the ovary with

The *stigma*, which is sometimes a mere viscid point, but more frequently an enlargement of the summit of the style, is variously shaped, being globular, flat, lobed, &c. If there be no style, the stigma is said to be sessile.

In the majority of flowers there is but one pistil; but very often there is a single ovary, which bears several styles and stigmas. In this case the ovary usually consists of several cells, each of which, with its style and stigma, is termed a *carpel ;* and the same name is given to each of the ovaries in such flowers as Marsh Marigold, vol. i. p. 15, where they are distinct; and in Blackberry, p. 192, where they are united.

Both calyx and corolla, it has been said above, may be absent. Not so with respect to stamens and pistils ; for, unless they are present, no seed can be perfected. It is not, however, essential that they should both be found in the same flower. Sometimes on the same plant flowers are to be found, some of which bear stamens only, others pistils only ; and not unfrequently these organs grow, not only in separate flowers, but on different plants. In either case, those flowers alone which contain pistils produce seeds, and are therefore termed *fertile ;* while those containing stamens only, are called *barren*. The external structure of barren and fertile flowers is often very dissimilar, as in Willow, vol. ii. p. 180 ; and Oak, p. 185. When the ovary is inserted above the base of the perianth, it is said to be *superior*, as in Crowfoot, vol. i. p. 10; when below, *inferior*, as in Rose, p. 199. In like manner the perianth is said to be superior or inferior, according as it is inserted above or below the ovary.

THE FRUIT.—As the flower withers, the ovary enlarges and becomes the *fruit*, that is, the seed, with its case or covering, also called a *pericarp*, (from *peri*, around, and *carpos*, fruit.) Among the various forms of fruit, the principal are—

The *capsule*, (from *cápsula*, a little box,) a dry case, either opening by *valves*, as in Pink, vol. i. p. 88 ; by *teeth*, as in Lychnis, vol. i. p. 93; by *pores*, as in Poppy, p. 26 ; or by splitting all round, as in Pimpernel, vol. ii. p. 133.

The *silique* and *silicle*, described at vol. i. p. 35.

The *pod*, or *legume*, a long seed-vessel, differing from the silique in having no partition, and bearing the seeds in a single row, as in the Pea and Bean Tribe, vol. i. p. 138.

The *berry*, a juicy or mealy fruit, bearing the seeds immersed in pulp, as in Elder, Currant, &c.

The *nut*, a dry fruit, composed of a hard shell, containing a seed, as in Hazel, vol. ii. p. 186 ; and Gromwell, p. 50.

The *drupe*, a nut enclosed in pulp, as the Plum and Cherry.

The *cone*, a collection of *imbricated* or overlapping scales, each of which covers two seeds, vol. ii. p. 188.

THE SEED.—A seed is said to be *dicotylédonous*, when
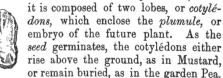
it is composed of two lobes, or *cotylédons*, which enclose the *plumule*, or embryo of the future plant. As the *seed* germinates, the cotylédons either rise above the ground, as in Mustard, or remain buried, as in the garden Pea. Plants bearing seeds of this structure compose the first Natural Class, DICOTYLÉDONOUS PLANTS, or EXOGENS, vol. i. p. 1. When the seed is not separable into two parts,

it is termed *monocotylédonous;* and plants bearing such seeds compose the Second Natural Class, MONOCOTYLÉ- DONOUS PLANTS, or ENDOGENS, vol. ii. p. 191.

RECEPTACLE.—This name is given to that part of the flower on which all the others rest. It is most conspi- cuous in the Compound Flowers, vol. i. p. 317, where it is sometimes *conical,* as in Daisy, p. 374 ; *chaffy,* as in Cat's-ear, p. 332 ; *bristly,* as in Thistle, p. 346 ; or *dotted,* as in Dandelion, p. 340.

NECTARY.—Any distinct organ in a flower which con- tains honey; for instance, the scale at the base of the petals in Crowfoot, vol. i. p. 8 ; the spurs of the Colum- bine, p. 18, &c.

INFLORESCENCE.—This term is used to denote the arrangement of flowers on the stem.

A flower-stalk springing directly from the root, and bearing no leaves, is termed a *scape,* as in Primrose, vol. ii. p. 129.

When it is inserted in the angle between the main stem and a leaf, it is termed *axillary,* as in Balsam, vol. i. p. 130.

When it is at the extremity of the main stem, having no leaves beyond it, it is said to be *terminal,* as in Grass of Parnassus, vol. i. p. 120.

A flower-stalk which bears but one flower, is said to be *simple,* as in Grass of Parnassus, vol. i. p. 120.

A stalk bearing a number of sessile flowers arranged one above another, is termed a *spike,* as in Plantain, vol. ii. p. 142.

When, instead of being sessile, the flowers are sup- ported on simple stalks, the inflorescence is a *cluster,* as in Melilot, vol. i. p. 150.

A *panicle* differs from a cluster in being branched, as in Spurrey, vol. i. p. 98.

A *corymb* differs from a cluster in bearing the lower flowers on long stalks, while the upper are sessile, or nearly so, as in Stock, vol. i. p. 65.

In a *cyme* the stalks are irregularly branched, but

the flowers are nearly level, as in Elder, vol. i. p. 295.

The *umbel* is a mode of inflorescence in which the flower-stalks spring from a common centre, and bear each a single flower, as in Ivy, vol. i. p. 285. When the stalks bear, instead of a single flower, a second umbel, the inflorescence is a *compound umbel,* the primary division being termed a *general umbel,* the secondary a *partial.* This mode of inflorescence is common in the Umbelliferous Tribe, vol. i. p. 244.

A *head* resembles a simple umbel, except that the flowers are all sessile, as in Scabious, vol. i. p. 316.

A *catkin* resembles a spike, except that the flowers are enclosed each within a scale-like bract, as in Hazel, vol. ii. p. 186.

Other terms which are employed in the body of the work will be explained as they occur, or in the description which precedes the summary of each Natural Order. A Glossary will also be found at the end of the second volume, containing definitions of most of the common terms in use.

CHAPTER II.

SYSTEMATIC BOTANY.

It has been already stated that the only parts of a plant which are necessary to the production of seed are the *stamens* and *pistils*. On these organs, therefore, the learned Linnæus fixed, when framing his artificial arrangement of plants, as affording the readiest means of referring to the written characters of plants described in his works.[1] He nowhere claims the honour of having discovered their importance; on the contrary, he expressly alludes to a popular opinion that the fact was known to Thomas Millington, a naturalist of the preceding century. But whoever discovered the fact, it lay idle and unnoticed until Linnæus invented and perfected the system founded on the fact. This can be proved, both by the jealousy with which it was received by the naturalists of the day, whose favourite methods have disappeared before it, as well as by the acrimony with which the name and works of Linnæus are assailed by some modern botanists—men who, while they disparage the works of their great leader, find it impossible to quit the track that he has trodden out for them, from a conviction that truth lies in the path that he has prescribed, and nowhere else.

The first division of Vegetables, according to the system of Linnæus, is into TWENTY-FOUR CLASSES, depending on the number, position, and relative proportion of the *Stamens*.

The first eleven CLASSES are characterised solely by the number of the stamens, and are thus named :—

(1) The number of species known to Linnæus in 1753, when he published his " Species Plantarum," amounted only to 7,300.

Class I. MONANDRIA, one stamen.
   II. DIANDRIA, two stamens.
   III. TRIANDRIA. three ditto.
   IV. TETRANDRIA, four ditto.
   V. PENTANDRIA, five ditto.
   VI. HEXANDRIA, six ditto.
   VII. HEPTANDRIA, seven ditto.
   VIII. OCTANDRIA, eight ditto.
   IX. ENNEANDRIA, nine ditto.
   X. DECANDRIA, ten ditto.
   XI. DODECANDRIA, twelve to nineteen ditto.

The Twelfth and Thirteenth Classes are distinguished by the situation as well as number of the stamens.

Class XII. ICOSANDRIA, twenty stamens, or more, inserted on the calyx.

   XIII. POLYANDRIA, twenty stamens, or more, inserted on the receptable.

The Fourteenth and Fifteenth Classes, by the number and relative proportion of the stamens.

Class XIV. DIDYNAMIA, four stamens, two longer than the others.

   XV. TETRADYNAMIA, six stamens, four long and two short.

The Classes from the Sixteenth to the Nineteenth are distinguished by the combination of the stamens with each other, no account being taken of their number.

Class XVI. MONADELPHIA, stamens all united by their filaments.

   XVII. DIADELPHIA, stamens combined by their filaments into two sets.

   XVIII. POLYADELPHIA, stamens combined by their filaments into three or more sets.

   XIX. SYNGENESIA, stamens united by their anthers; flowers compound.

In *Class* XX. GYNANDRIA, the stamens are inserted on the pistil, and so combined with it as to form a central *column*.

The Twenty-first, Twenty-second, and Twenty-third

Classes are distinguished by the stamens and pistils being in separate flowers.

Class XXI. Monœcia, stamens and pistils in different flowers, but on the same plant.

XXII. Diœcia, stamens and pistils in different flowers and on separate plants.

XXIII. Polygamia, flowers of three kinds; some having stamens only, some pistils only, and some both stamens and pistils.

XXIV. Cryptogamia, flowerless plants, or those in which stamens and pistils have not been detected, fructification being performed by other organs.

Each of the Classes is subdivided into two or more Orders. The Orders of the first Thirteen Classes depend on the number of the *Pistils*. Thus, any plant in either of the classes from Monandria to Polyandria must be placed in one or other of the following Orders :—

Order Monogynia, one pistil.
Digynia, two pistils.
Trigynia, three ditto.
Tetragynia, four ditto.
Pentagynia, five ditto.
Hexagynia, six ditto.
Heptagynia, seven ditto.
Decagynia, ten ditto.
Dodecagynia, twelve ditto.
Polygynia, many ditto.

In the Fourteenth Class, Didynamia, there are two Orders.

Order I. Gymnospermia, ovaries four, one-seeded.
II. Angiospermia, ovary single, many-seeded.

In the Fifteenth Class, Tetradynamia, there are two Orders.

Order I. Siliculosa, fruit a *silicle*, or short pod without a partition.
II. Siliquosa, fruit a *silique*, or long pod without a partition.

The Orders of the Sixteenth, Seventeenth, and Eighteenth Classes, MONADELPHIA, DIADELPHIA, and POLYADELPHIA, depend on the *number* of the *stamens :* thus, *Class* MONADELPHIA, *Order* PENTANDRIA, includes plants having *five* stamens united by their filaments into one set ; *Class* DIADELPHIA, *Order* DECANDRIA, plants having *ten* stamens combined by their filaments into *two* sets ; and *Class* POLYADELPHIA, *Order* POLYANDRIA, plants with more than *twenty* stamens combined by their filaments into *three* or *more* sets.

In the Nineteenth Class, SYNGENESIA, the Orders depend on the structure and arrangement of the *florets ;* but as they are Orders nearly identical with the *Groups* into which the COMPOUND FLOWERS are distributed, their limits need not be assigned here.

In the Twentieth, Twenty-first, and Twenty-second Classes, GYNANDRIA, MONŒCIA, and DIŒCIA, the Orders are determined by the *number* of the stamens. Plants, for example, having *one* stamen are in the *Order* MONANDRIA; those with *two* stamens, in the *Order* DIANDRIA, &c.

The Twenty-third Class, POLYGAMIA, contains only one British *Order*, namely MONŒCIA, in which there are three different kinds of flowers—those with stamens only, those with pistils only, and those with both stamens and pistils *on the same plant.*

As the limits of these volumes exclude all mention of the plants in the extensive *Class* CRYPTOGAMIA, it is not necessary to enumerate the *Orders* into which it is divided.[1]

(1) The student is recommended to commit to memory the names, rather than the numbers of the Classes and Orders. While he does this he will find it useful to bear in mind that the names both of Classes and Orders are of Greek etymology, and that the prefixes are mostly numerals. Thus, *mon* signifies one; *di*, two; *tri*. three; *tetra*, four; *pent*, five; *hex*, six; *hepta*, seven; *oct*, eight; *enne*, nine; *dec*, ten; *dodec*, twelve; *icos*, twenty; and *poly*, many. The root *dynam* signifies power or excess: thus, *Didynamia* means "the excess of two ;" *Tetradynamia*, "the excess of four." *Adelphia* signifies a brotherhood, as *Monadelphia*, " one brotherhood," or united set of stamens; *Syngenesia* signifies "a growing together," in allusion to the combination of the anthers in that class. The termination *œcia* denotes "a house-

This arrangement brings together, for no other purpose than for convenience of reference, plants dissimilar in stucture, habit, and properties. It is, therefore, an Artificial System ; as such Linnæus proposed it, and such he always professed it to be. " I have never pretended that the method was natural," he says, in his letter to Haller. " A Natural System," he repeatedly remarks, in his other writings, " is the first and last object to be aimed at by botanists.[1] .... A perfect system of this kind should assemble plants allied in habit, mode of growth, properties, and uses." Of such a system he left a slight sketch ; but the rich store of plants which has been laid open to modern botanists never came within his reach; it is, therefore, not surprising that, being well aware of his defective materials, he never attempted to fill the sketch in. Make it as complete as he would, in a few years it would have been imperfect and useless. Not so, however, his Artificial System, which, still marked by the limits that he assigned, not only offers facilities for forming an acquaintance with the names of plants, but affords ready means of reference to any System in which plants are arranged according to their natural characters. It is not, therefore, too much to say that the Artificial System of Linnæus has served a double purpose. Before a Natural Method was arranged, it was the only one that was available ; and now that it is

hold ;" so in the Class *Monœcia*, the stamens and pistils may be supposed to occupy separate apartments in one house. *Polygamia* signifies " many kinds of fructification ;" *Cryptogamia*, " concealed fructification." *Andria* denotes stamens; *gynia*, pistils : thus *Triandria* includes flowers with *three stamens* ; *Digynia* flowers with *two pistils;* and *Gynandria* flowers with *pistils and stamens united.* In the two Orders of the Class *Didynamia*, the term *Gymnospermia* denotes " naked seed;" the fruit being apparently destitute of a covering: *Angiospermia* implies that the seeds are enclosed in a " seed-vessel." The terms *Siliculosa* and *Siliquosa* are explained in the text.

(1) Methodi Naturalis fragmenta inquirenda sunt. Primum et ultimum hoc in Botanicis desideratum est. Plantæ omnes utrinque affinitatem monstrant uti territorium in mappâ geographicâ.—*Lin. Phil. Bot.* Aph. 77.

Methodus Naturalis est ultimus finis Botanices.—*Ibid.* Aph. 163.

Naturalis Character ab omni Botanico teneatur oportet.—*Ibid.* Aph. 191.

Classes, quo magis naturales, eo, cæteris paribus, præstantiores sunt. Adfines conveniunt habitu, nascendi modo, proprietatibus, viribus, usu. Summorum Botanicorum hodiernus labor in his sudat, et desudare decet; methodus Naturalis hinc ultimus finis Botanices est et erit.—*Ibid.* Aph. 206.

superseded, it is still eminently useful as an index, or a catalogue of the contents of its successor; the secondary use being one which, perhaps, Linnæus himself scarcely contemplated.

It is not necessary here to give an account of the various Natural Systems which have been proposed. Suffice it to say, the one generally adopted in Britain is a modification of those of Jussieu and De Candolle. Here the whole Vegetable Kingdom is divided into three great CLASSES.

CLASS I.   DICOTYLÉDONES.

In this Class are placed such plants as produce seeds divisible into two lobes or *cotyledons* (vol. i. p. 1).   It is subdivided into Four *Sub-classes*, THALAMIFLORÆ, CA-LYCIFLORÆ, COROLLIFLORÆ, and MONOCHLAMYDEÆ.

Sub-class I.   THALAMIFLORÆ.

Flowers furnished with calyx and corolla; *petals* distinct, inserted into the receptacle, or *thalamus; stamens* springing from the base of the *ovary.*—This Sub-class contains Twenty-two British Orders.   (Vol. i. pp. 2—132.)

Sub-class II.   CALYCIFLORÆ.

Flowers furnished with calyx and corolla; *sepals* distinct, or united; *petals* distinct; *stamens* inserted in the *calyx,* or close to its base.—This Sub-class contains Eighteen British Orders, numbered from XXIII. to XL.   (Vol. i. pp. 133—290.)

Sub-class III.   COROLLIFLORÆ.

Flowers furnished with calyx and corolla; *petals* united, bearing the stamens.—In this Sub-class there are twenty-seven British Orders, numbered from XLI. to LXVII.   (Vol. i. p. 290, to vol. ii. p. 143.)

Sub-class IV. MONOCHLAMYDEÆ.[1]

Perianth single, or none. This Sub-class contains thirteen British Orders, numbered from LXVIII. to LXXIX. (Vol. ii. pp. 144—190.)

CLASS II. MONOCOTYLEDONES.

Seeds with a single *cotyledon* (vol. ii. p. 191). It is subdivided into Two *Sub-classes*, PETALOIDEÆ and GLU-MACEÆ.

Sub-class I. PETALOIDEÆ.

Flowers with petals.—This Sub-class contains sixteen British Orders, numbered from LXXX. to XCVI. (Vol. ii. p. 192 to the end.)

Sub-class II. GLUMACEÆ.

Flowers formed of chaffy scales, or *glumes*.—This Sub-class contains the Grasses and Sedges.

CLASS III. ACOTYLEDONES.

Flowerless plants. Here are placed the Ferns, Mosses, Liverworts, Lichens, Sea-weeds, and Fungi.

Each of the *Natural Orders*, or *Tribes*, alluded to above, consists of a number of plants which are more or less like one another in various respects, especially in the organs of fructification. The plants comprised in each Tribe are again distributed into *genera*, or *families*, each genus including all plants which resemble one another yet more closely in the essential characters of fructification. A *species*, or *kind*, is an assemblage of individual plants agreeing with each other in *all* essential points ; and individuals which differ one from

(1) From the Greek *monos*, one, and *chlamys*, a mantle or covering ; the plants of this Sub-class never having both calyx and corolla.

another in minor points, such as an irregular formation
of leaves or mode of growth, unusual colour of flowers,
extraordinary number of petals, &c., are termed *varieties*.
These words are frequently used loosely in common con-
versation, but the habit cannot be too carefully avoided
in botanical descriptions, as calculated to produce great
confusion. Throughout these pages they will be employed exclusively with the meanings above assigned,
which will be rendered clearer by the following examples :—The wild sweet-scented Violet is called by botanists *Viola odoráta ;* the former name, *Viola*, indicating
that it belongs to the *genus* so called, and being, therefore, termed its *generic* name. Besides the scented Violet,
we have in England the Dog-Violet, the Marsh-Violet,
the Pansy, and several others, all belonging to the same
genus, and, therefore, described under the name *Viola*.
But the Dog-Violet differs from the Sweet-scented, in
having acute sepals and leafy stems, whereas the latter
has blunt sepals, and the leaves spring directly from the
roots. The Dog-Violet is, therefore, a distinct *species*,
Viola canína. The Marsh-Violet and Pansy differ also
in important characters ; they are, therefore, also considered distinct species, the fact being indicated by the
addition of the *specific* or *trivial*[1] names, *palustris* and
tricolor, to the *generic* name, *Viola*. The flowers of the
scented *Violet* are sometimes white and sometimes blue ;
garden specimens are often tinged with pink, and still
more frequently, double. These characters being either
unimportant, or inconstant—for blue flowers generally
have a great tendency to sport to white, and double
flowers are not perpetuated by seed—the blue, white,
pink, and double sweet Violets are not considered distinct species, but mere *varieties*. Now there are many

(1) No little merit is due to Linnæus for inventing the specific or trivial
name of plants. The method in use previously to his time was to attach to
every plant some such title as the following : *Gramen Xeɩampelinum, Mi-
liacea, prætenuis ramosaque sparsa panicula, sire, Xerampelino congener,
arvense, æstivum; gramen minutissimo semine.* The name of this grass Linnæus expressed with accuracy and simplicity by the two words, *Poa bulbosa.*

plants which bear a close resemblance to a Violet in the structure of their flowers and seeds, but yet differ so far that they cannot be reduced under the same *genus ;* they are, therefore, placed with it in the same *Tribe,* called VIOLACEÆ, all the genera in which differ in essential points from the genera which compose other Tribes, but agree with a vast number in having *two-lobed seeds* and *leaves with netted veins,* two of the characters of DICOTYLEDONOUS PLANTS. In this Class it is arranged with plants furnished with both calyx and corolla, and having their petals distinct and inserted with the stamens into the receptacle.

The plant of which we have been speaking belongs, then, to the—

CLASS I. DICOTYLÉDONES.

SUB-CLASS I. THÁLAMIFLORÆ.

Order or *Tribe* IX. VIOLACEÆ.

Genus 1, Viola.

Species 2, *odoráta.*

Variety, blue, white, or *double.*

In the Linnæan system the same plant is placed in the *Class* PENTANDRIA, which comprises flowers having *five stamens ;* and in the *Order* MONOGYNIA, which includes such of them as have *one pistil.*

In this work the British genera and species are arranged in their Natural Orders or Tribes ; and a synopsis is also given of the genera only, as they stand in the Linnæan Classes and Orders. (Vol. i. p. xxviii. &c.)

The student, it is presumed, wishes to determine the genus and species, or to find the name, of the plants which he meets with in his walks. Suppose him to have found a small shrubby plant with oblong leaves and large yellow flowers : in which volume, and in what part of it, must he look for a description which he may compare with the specimen in his hand ?   On examining a flower (he will do well to select one which is just expanded), he will discover a large number of *stamens*, evidently more than twenty, inserted in the receptacle,

HELIANTHEMUM VULGARE.

and he will have no doubt that it belongs to the Linnæan *Class* POLYANDRIA.   In the centre of the stamens he will detect a single *pistil*, and hence will conclude that it should be referred to the *Order* MONOGYNIA in that Class.   He will, accordingly, turn to *Class* POLY-

ANDRIA, *Order* MONOGYNIA (vol. i. p. xlvii.), and will proceed to read the characters of the genera there described,
comparing such descriptions with his specimen. Opposite
the first genus, *Delphínium,* he will find the characters,
" *Sepals* 5, coloured, the upper one spurred." This is
no description of his plant, so he passes on to 2. *Papáver :*
"*Sepals* 2," &c. He reads no further, but passes on to
6. *Actæa :* "*Sepals* 4." Neither will this do, for his
plant has 5 *sepals,* of which the two outer are much
smaller than the rest. The next, genus 6. *Heliánthemum,*
has " *Sepals* 5, the two outer smallest, or sometimes
wanting ; *petals* 5." To this genus, then, his plant belongs ;—the short description which follows also suits
his plant : " Small shrubs with oblong leaves, and showy,
white or yellow flowers." He is then referred to the
ROCK-ROSE TRIBE (vol. i. p. 72), where are enumerated
the leading characters of the Tribe and of all the British
genera included in it. Being assured, by a perusal of
these, that his plant is a *Heliánthemum,* he proceeds to
ascertain its *species,* and finally satisfies himself that he
holds in his hand a specimen of a plant known among
botanists by the name of *Heliánthemum vulgáre* (Common Rock-Rose.)

To take another example. The plant of which it is
desired to discover the name is an erect, herbaceous
plant, with smooth leaves and long spiked clusters of
green flowers. The number of stamens here is variable,
so that there is some doubt whether it should be referred
to the *Class* DODECANDRIA, including plants with from
12 to 20 stamens, or to the *Class* POLYANDRIA, with
more than 20. About the number of *pistils,* however,
there can be no doubt—they are distinctly 3. Accordingly, if we turn to *Class* DODECANDRIA, *Order* TRYGYNIA
(p. xlvi.), we find only the genus " *Reséda.* Herbaceous
or somewhat shrubby plants with furrowed stems, smooth
leaves, and terminal spikes or clusters of greenish flowers."
This description is accompanied by a reference to the
ROCKET TRIBE (vol. i. p. 71). If, on the other hand, we

refer to the *Class* POLYANDRIA, we find the *Order* PENTA-
GYNIA, including plants with from 2 to 6 pistils.   Here,
mention of *Reséda* occurs again, and other characters
are given, equally descriptive of the plant in question,

RESEDA LUTEOLA.

namely : " *Flowers* irregular ; *capsule* solitary, open at
the top."   We are again referred to the ROCKET TRIBE
(vol. i. p. 71), where we find a fuller description of the
tribe and genus, followed by such a description of the
only two British species as assures us that our plant is
R. Lutéola, or Dyer's Rocket.

One more example will suffice.   The plant to be
examined is a small herb with somewhat fleshy leaves

and white flowers, with 6 stamens, four long and two short. Had they all been of equal length we should have had no hesitation in referring the plant to the *Class* HEXANDRIA; but as this is not the case, we turn to the *Class* TETRADYNAMIA (vol. i. p. xlix). Here we find that " all the plants in this Class belong to the CRUCIFEROUS TRIBE, vol. i. p. 35." On turning to the reference, we learn that the division of this TRIBE corresponds with that of the *Linnæan Class*,—namely, into plants bearing *silicles* and those bearing *siliques*. The plant in question bears a *silicle*, or *pouch*, and must be searched for in the group " † *Pouch 2-valved, with a central vertical partition.*" *Thlaspi* and *Capsella* are passed over, as having *flat* pouches; *Hutchinsia* and *Teesdalia* have a *keeled* pouch, the cells of which are 2-seeded; in *Lepidium* the valves of the pouch are keeled, and the cells 1-seeded; but in *Cochlearia* the pouch is globose, or nearly so, the valves are not flattened, and the seeds are numerous. Our plant, then, is a *Cochlearia*; and on turning to genus 6 (p. 45), we find that our specimen, which has an egg-shaped pouch and triangular leaves, is *C. Dánica* (Danish Scurvy-Grass).

LINNÆAN ARRANGEMENT.

Class I. MONANDRIA. 1 *Stamen.*

Order I. MONOGYNIA. 1 *Pistil.*

1. SALICORNIA. *Calyx* fleshy ; *petals* 0 ; *stigma* 2-3 cleft.—Sea-side plants with jointed stems, inconspicuous flowers, and no true leaves. GOOSE-FOOT TRIBE. Vol. ii. p. 144.

2. HIPPÚRIS. *Calyx* minute ; *petals* 0 ; *stigma* undivided.—Aquatic plants with whorled leaves, and inconspicuous axillary flowers. MARE'S-TAIL TRIBE. Vol. i. p. 212.

3. ZOSTÉRA. *Calyx* 0 ; *petals* 0 ; *stamens* and *pistils* arranged alternately in 2 rows.—Marine plants, with long grass-like submersed leaves. POND-WEED TRIBE. Vol. ii. p. 254.

4. VALERIANA. *Calyx* bearing a feathery appendage to the pericarp ; *corolla* of 1 *petal* spurred at the base. —Herbaceous plants with opposite leaves and small flowers. VALERIAN TRIBE. Vol. i. p. 308.

5. ALCHEMILLA. *Calyx* 8-cleft, bearing the *stamens,* which vary in number from 1 to 4 ; *petals* 0.—Herbaceous plants with stipuled leaves and small green flowers. —ROSE TRIBE. Vol. i. p. 170.

Order II. DIGYNIA. 2 *Pistils.*

6. CALLÍTRICHÉ. Small aquatic plants with opposite leaves, inconspicuous axillary flowers without petals, and 4-seeded fruit. MARE'S-TAIL TRIBE. Vol. i. p. 212.

CLASS II.　DIANDRIA.　2 *Stamens.*

ORDER I.　MONOGYNIA.　1 *Pistil.*

1. LIGUSTRUM.　*Corolla* 4-cleft, regular ; *fruit* a berry.—A shrub with smooth leaves, and white panicled flowers.　OLIVE TRIBE.　Vol. ii. p. 25.

2. VERÓNICA.　*Corolla* unequally 4-cleft ; *fruit* a 2-celled capsule.—Herbaceous plants with opposite leaves and blue or flesh-coloured flowers.　FIG-WORT TRIBE. Vol. ii. p. 70.

3. PINGUÍCULA.
4. UTRICULARIA.
}
Corolla 2-lipped and spurred ; *fruit* a 1-celled capsule.　Aquatic or bog plants with delicate yellow, pink, or purple flowers.　BUTTER-WORT TRIBE.　Vol. ii. p. 123.

5. LÝCOPUS.
6. SÁLVIA.
}
Corolla gaping ; *fruit* 4-lobed and 4-seeded. — Herbaceous plants with square stems and opposite leaves.　LA-BIATE TRIBE.　Vol. ii. p. 91.

7. CIRCÆA.　*Sepals* 2 ; *petals* 2.—Herbaceous plants with opposite leaves, and small pinkish white flowers. WILLOW-HERB TRIBE.　Vol. i. p. 206.

8. FRÁXINUS.　*Calyx* 0 ; *corolla* 0 ; *fruit* winged, 1-seeded.—A lofty tree.　OLIVE TRIBE.　Vol. ii. p. 25.

9. LEMNA.　Minute, stemless, aquatic, floating plants, consisting entirely of a few simple roots and small leaves, embedded in which last, flowers are produced, but rarely. Nat. Ord. DUCK-WEED TRIBE.　Vol. ii. p. 253.

CLASS III.　TRIANDRIA.　3 *Stamens.*

ORDER I.　MONOGYNIA.　1 *Pistil.*

1. VALERIANA.
2. FÉDIA.
}
Calyx crowning the fruit ; *corolla* 5-cleft — Herbaceous plants with opposite leaves and small flowers. VALERIAN TRIBE.　Vol. i. p. 308.

            ⎧ *Perianth* 6-cleft superior.—Her-
3. IRIS.   ⎪ baceous plants with bulbous, or
4. CROCUS. ⎨ tuberous roots, sword-shaped, or
5. TRICHONÉMA. ⎪ grass-like leaves, and beautiful
            ⎩ flowers. IRIS TRIBE. Vol. ii. p. 205.
6. JUNCUS. *Perianth*, 6-cleft, inferior.—Mostly marsh
plants with cylindrical stems and leaves, and brownish
green flowers. RUSH TRIBE. Vol. ii. p. 232.

* Many of the SEDGE TRIBE belong to this Class and
Order ; their inflorescence is in terminal spikes formed
of chaffy scales, or *glumes*, with a flower at the base of
each. Not described in these volumes.

ORDER II. DIGYNIA. 2 *Pistils.*

This Order contains only the GRASS TRIBE, the flowers
of which consist of 2 chaffy valves or glumes arranged in
small spikes (*spikelets*), each of which has a pair of
glumes at the base, outside. The Grasses and Sedges
are not described in these volumes.

ORDER III. TRIGYNIA. 3 *Pistils.*

7. MONTIA. *Sepals* 2 ; *capsule* 3-valved.—A small
aquatic plant, with opposite leaves and minute white
flowers. PURSLANE TRIBE. Vol. i. p. 225.
8. HOLÓSTEUM. *Sepals* 5 ; *capsule* opening with 6
teeth.—A small plant with opposite leaves and white
flowers which grow in umbels. CHICKWEED TRIBE.
Vol. i. p. 84.
9. POLYCARPON. *Sepals* 5 ; *capsule* 3-valved.—A
small plant with white flowers, bearing its lower leaves
in whorls of 4. KNOT-GRASS TRIBE. Vol. i. p. 226.
10. TILLÆA. *Sepals* 3 ; *carpels* 3.—A minute plant
with fleshy leaves and greenish-white flowers. STONE-
CROP TRIBE. Vol. i. p. 230.

29. VINCA. *Corolla* salver-shaped with oblique segments.—Somewhat shrubby plants with blue flowers, and bearing their seeds in 2 erect capsules, which do not burst. PERIWINKLE TRIBE. Vol. ii. p. 30.

*** *Perianth double; corolla of 1 petal; ovary inferior.*

30. JASÍONÉ. *Flowers* in heads; *anthers* united at the base.—An herbaceous plant with rough leaves and heads of blue flowers surrounded, like the COMPOUND FLOWERS, by an involucre. BELL-FLOWER TRIBE. Vol. ii. p. 1.

31. LOBELIA. *Corolla* 2-lipped, split to the base on the upper side; *anthers* united.—Herbaceous plants with showy flowers, of which the lower lip is deeply lobed. LOBELIA TRIBE. Vol. ii. p. 6.

32. CAMPÁNULA. ⎰ *Corolla* bell- or wheel-shaped;
33. PHYTEUMA. ⎱ *style* downy on its upper part.— Herbaceous plants, mostly with handsome blue flowers. BELL-FLOWER TRIBE. Vol. ii. p. 1.

34. LONICÉRA. *Corolla* with a long tube, gaping; *berry* crowned by the calyx.—Mostly twining shrubs with fragrant flowers, which are either axillary or grow in heads. WOODBINE TRIBE. Vol. i. p. 293.

**** *Perianth double; corolla of 4—5 petals; ovary superior.*

35. RHAMNUS. *Stamens* opposite the petals; *fruit* a 2—4 seeded berry.—Small trees with greenish flowers and black berries. BUCKTHORN TRIBE. Vol. i. p. 135.

36. EUÓNYMUS. *Stamens* alternate with the petals; *fruit* an angular capsule.—A shrub remarkable for its green branches and flowers, and bright rose-coloured capsules. SPINDLE-TREE TRIBE. Vol. i. p. 133.

37. IMPATIENS. *Sepals* 2, coloured; *corolla* irregular, spurred.—Succulent herbaceous plants, remarkable for the curious structure of their flowers and elastic capsules. BALSAM TRIBE. Vol. i. p. 129.

38. Vióla.   *Sepals 5 ; corolla* irregular, spurred.—
Herbaceous plants with blue, white, yellow, or variegated
flowers, which are often fragrant.   Violet Tribe.   Vol.
i. p. 74.

***** *Perianth double; corolla of 5 petals ; ovary
inferior.*

39. Ribes.   *Calyx* 5-cleft, bearing the petals and
stamens ; *berry* juicy.—Shrubs with green flowers and
eatable fruit.   Gooseberry and Currant Tribe.   Vol. i.
p. 237.

40. Hédera.   *Calyx* of 5 teeth ; *petals* and *stamens*
inserted on the ovary ; *berry* mealy.—An evergreen
climbing shrub, bearing its green flowers in simple
umbels.   Ivy Tribe.   Vol. i. p. 284.

****** *Perianth single.*

41. Glaux.   *Corolla* bell-shaped, 5-lobed ; *capsule*
with about 5 seeds.—A sea-side plant, with succulent
leaves and pink axillary flowers.   Primrose Tribe.
Vol. ii. p. 127.

42. Illécebrum.   *Sepals* 5, awl-shaped, white ; *seed* 1.
—A rare aquatic, with slender tangled stems, minute
leaves, and whorled flowers.   Knot-grass Tribe.   Vol. i.
p. 226.

43. Thesium.   *Perianth* 4—5-cleft ; *fruit* a *drupe*
crowned by the calyx.—A parasitic plant, with branched
herbaceous stems and terminal clusters of whitish flowers.
Sandal-wood Tribe.   Vol. ii. p. 161.

Order II.   DIGYNIA.   *2 Pistils.*

44. Hydrocótylé, to 66. Myrrhis.   *Petals* 5 ; seeds
2, crowned by the calyx.—Herbaceous plants, mostly
with hollow, jointed stems, compound leaves, and umbels
of small flowers.   Umbelliferous Tribe.   Vol. i. p. 244.

67. Gentiana.   *Corolla* of 1 petal ; *capsule* 1-celled,

2-valved, many-seeded. — Bitter herbs with opposite leaves and tubular flowers. GENTIAN TRIBE. Vol. ii. p. 31.

68. CÚSCUTA. *Corolla* of 1 petal ; *capsule* 4-seeded. —Leafless parasitic plants, with red tangled stems, and wax-like flowers. BINDWEED TRIBE. Vol. ii. p. 41.

69. HERNIARIA. *Sepals* 5 ; *petals* or *barren stamens* 5, alternate with the stamens.—A small, somewhat shrubby plant, with prostrate stems and numerous small green flowers. KNOT-GRASS TRIBE. Vol. i. p. 226.

70. SCLERANTHUS. *Calyx* 5-cleft, contracted at the mouth of the tube ; *petals* 0.—A low, repeatedly forked, herbaceous plant, with linear leaves and small green flowers. KNOT-GRASS TRIBE. Vol. i. p. 226.

71. CHENOPODIUM.
72. BETA.
73. SÁLSOLA.

{ *Sepals* 5, sometimes combined at the base ; *petals* 0 ; *seed* 1, enclosed by the calyx.—Herbaceous or somewhat shrubby plants, with fleshy leaves and green flowers. GOOSE-FOOT TRIBE. Vol. ii. p. 144.

74. POLÝGONUM. *Perianth* single, coloured ; *seed* 1, enclosed in the perianth, not winged.— Herbaceous plants, with alternate leaves and sheathing stipules. PERSICARIA TRIBE. Vol. ii. p. 151.

75. ULMUS. *Perianth* single, coloured ; *seed* 1, winged all round.—Lofty trees with purplish flowers, which appear before the rough leaves. ELM TRIBE. Vol. ii. p. 177.

ORDER III. TRIGYNIA. 3 *Pistils*.

* *Perianth double*.

76. VIBURNUM.
77. SAMBÚCUS.

{ *Corolla* of 1 petal, superior ; *fruit* a berry.—Shrubs with white flowers growing in cymes. WOODBINE TRIBE. Vol. i. p. 293.

78. TÁMARIX. *Stigmas* sessile, feathery ; *seeds* tufted with down.—A shrub or small tree, with long, flexible

branches, scale-like leaves, and rose-coloured flowers.
TAMARISK TRIBE. Vol. i. p. 220.

79. CORRIGÍOLA. *Seed* 1.—A small herbaceous plant,
with prostrate stems, strap-shaped leaves, and tufted
white flowers. KNOT-GRASS TRIBE. Vol. i. p. 226.

80. STELLARIA.
81. HOLOSTEUM. } *Seeds* in a capsule, not tufted with down.— Small herbaceous plants, with opposite leaves and white flowers. CHICKWEED TRIBE. Vol. i. p. 86.

** *Perianth single.*

82. CHENOPODIUM. *Seed* 1.—Herbaceous or somewhat
shrubby plants, with fleshy leaves and green flowers.
GOOSE-FOOT TRIBE. Vol. ii. p. 144.

ORDER IV. TETRAGYNIA. 4 *Pistils.*

83. PARNASSIA.— An erect herbaceous plant, with
handsome, solitary, cream-coloured flowers, beautifully
veined. ST. JOHN'S WORT TRIBE. Vol. i. p. 116.

ORDER V. PENTAGYNIA. 5 *Pistils.*

84. STÁTICÉ. *Petals* bearing the stamens.—Sea-side
plants with handsome flowers, remarkable for the membranaceous or chaffy character of the calyx. THRIFT
TRIBE. Vol. ii. p. 138.

85. LINUM. *Petals* alternate with the stamens; *capsule* 10-celled.—Slender plants with narrow leaves and
ornamental flowers, the petals of which soon fall off.
FLAX TRIBE. Vol. i. p. 106.

86. CERASTIUM.
87. SPÉRGULA. } *Capsule* 1-celled, many-seeded.— Herbaceous plants with opposite leaves and white flowers. CHICK-WEED TRIBE. Vol. i. p. 86.

88. SIBBALDIA. *Petals* and *stamens* inserted on the
calyx.—A humble alpine plant, with ternate leaves and

yellowish flowers, of which the stamens and pistils are very variable in number. ROSE TRIBE. Vol. i. p. 170.

ORDER VI. HEXAGYNIA. 6 *Pistils*.

89. DRÓSERA.—Bog plants of humble growth, remarkable for having their leaves clothed with red, viscid hairs. SUN-DEW TRIBE. Vol. i. p. 78.

ORDER VII. POLYGYNIA. *Many Pistils*.

90. MYOSÚRUS.—A small plant with narrow leaves and minute flowers, the pistils of which are crowded on a common receptacle, so as to resemble a mouse's tail. CROWFOOT TRIBE. Vol. i. p. 2.

CLASS VI. HEXANDRIA. 6 *Stamens of equal length*.

ORDER I. MONOGYNIA. 1 *Pistil*.

* *Perianth double*.

1. BÉRBERIS. *Sepals* 6, soon falling off; *fruit* a berry.—A thorny shrub with drooping clusters of yellow flowers. BARBERRY TRIBE. Vol. i. p. 21.

2. FRANKENIA. *Calyx* 6-cleft.—A sea-side plant with prostrate, wiry stems, linear leaves, and axillary, rose-coloured flowers. SEA-HEATH TRIBE. Vol. i. p. 82.

3. PEPLIS.
4. LYTHRUM.
Calyx in 12 divisions, 6 alternately smaller. — Herbaceous marsh plants, with opposite leaves and 4-cornered stems. LOOSESTRIFE TRIBE. Vol. i. p. 217.

** *Perianth single, superior*.

5. LEUCÓJUM.
6. GALANTHUS.
7. NARCISSUS.
Plants with bulbous roots and showy flowers, which before expansion are enclosed in a sheath. AMARYLLIS TRIBE. Vol. ii. p. 209.

*** *Perianth single, inferior.*

8. Aspáragus.   9. Convallaria.   10. Hyacinthus.
11. Allium.   12. Scilla.   13. Ornithógalum.
14. Gágea.   15. Túlipa.   16. Fritillaria. *Perianth*
of 6 petals, or 6-cleft; *anthers* bursting inwards.—Herbaceous plants often with bulbous roots, bearing showy or
(in Aspáragus) small green flowers. Lily Tribe. Vol.
ii. p. 218.

17. Juncus. 18. Lúzula. 19. Narthecium.	*Sepals* or *petals* 6, chaffy ; *fila-* *ments* downy. — Mostly marsh plants with pithy or jointed leaves and greenish flowers ; but Nar- thecium has yellow flowers and sword-shaped leaves, and Lúzula hairy, grass-like leaves. Rush Tribe. Vol. ii. p. 232.

20. Ácorus. *Flowers* sessile on a long central
column or *spadix.*—An aquatic, reed-like plant, the
leaves of which are fragrant when crushed. Cuckoo-
pint Tribe. Vol. ii. p. 249.

21. Polýgonum. *Perianth* 5-cleft ; *seed* 1.—Herbaceous plants with alternate leaves and sheathing stipules.
Persicaria Tribe. Vol. ii. p. 151.

Order II. DIGYNIA. 2 *Pistils.*

22. Oxýria.—An alpine, herbaceous plant, with kidney-shaped, juicy leaves and greenish flowers : whole
plant acid. Persicaria Tribe. Vol. ii. p. 151.

Order III. TRIGYNIA. 3 *Pistils.*

23. Rumex. *Seed* 1, 3-cornered.—Herbaceous plants
with astringent leaves, which have sheathing stipules,
reddish stems, and panicled, green flowers. Persicaria
Tribe. Vol. ii. p. 151.

24. Tofieldia. *Flowers* with a 3-lobed *bract* at the
base of each stalk.—A humble mountain-bog plant, with
sword-shaped leaves and yellowish, spiked flowers. Meadow-saffron Tribe. Vol. ii. p. 230.

25. Triglóchin. *Sepals* 6 ; *carpels* 3—6, 1-seeded. —Sea-side or marsh plants, with linear fleshy leaves, and slender spikes of inconspicuous green flowers. Arrow-grass Tribe. Vol. ii. p. 244.

26. Cólchicum. *Perianth* funnel-shaped, with a very long tube, coloured.—A bulbous-rooted plant, with broad leaves which wither before the flowers appear, and large, handsome, light-purple flowers. Meadow-saffron Tribe. Vol. ii. p. 230.

27. Elátiné. *Sepals* 3 ; *petals* 3 ; *capsule* 3-valved. —A minute aquatic plant, with rooting stems and opposite leaves. Water-wort Tribe. Vol. i. p. 83.

Order IV. HEXAGYNIA. 6 *Pistils.*

28. Actínocarpus.—An herbaceous aquatic, remarkable for the star-like arrangement of its carpels. Water-plantain Tribe. Vol. ii. p. 241.

Order V. POLYGYNIA. *Many Pistils.*

29. Alisma.—Herbaceous aquatics with smooth leaves and lilac flowers, each containing 3 sepals and 3 petals. Water-plantain Tribe. Vol. ii. p. 241.

Class VII. HEPTANDRIA. 7 *Stamens.*

Order I. MONOGYNIA. 1 *Pistil.*

1. Trientalis.—A northern herbaceous plant, with oblong, smooth leaves, and several terminal white flowers on slender stalks. Primrose Tribe. Vol. ii. p. 127.

Class VIII. OCTANDRIA. 8 *Stamens.*

Order I. MONOGYNIA. 1 *Pistil.*

* *Perianth double, inferior.*

1. Acer. *Capsules* 2, each furnished with a long wing.—Trees with lobed leaves and green flowers, which

expand at the same time with the leaves. Maple Tribe. Vol. i. p. 121.

2. Chlora.   *Calyx* 8-cleft; *corolla* 8-lobed.—An erect herbaceous plant, remarkable for the glaucous hue of its leaves, and for its yellow flowers, which expand only in fine weather. Gentian Tribe. Vol. ii. p. 31.

3. Monótropa.   *Flowers* in a cluster, the terminal one with 5 *sepals*, the lateral ones with 4.—A parasitic plant, consisting of a juicy stalk without leaves, and a drooping cluster of brownish flowers.— Bird's-nest Tribe. Vol. ii. p. 20.

4. Eríca.
5. Callúna.
6. Menziésia.
{ *Calyx* of 4 *sepals* or 4-cleft; *corolla* of 1 *petal*, 4-cleft.—Shrubs with evergreen, often rigid, opposite, or whorled leaves, and beautiful flowers. Heath Tribe. Vol. ii. p. 11.

* * *Perianth double, superior.*

7. Vaccínium.   *Corolla* of 1 petal, 4-cleft.—Small shrubs with simple, alternate leaves, and bell- or wheel-shaped flowers. Cranberry Tribe. Vol. ii. p. 8.

8. Œnothéra.
9. Epilóbium.
{ *Petals* 4.—Herbaceous plants with purple or yellow flowers and 4-valved capsules. Willow - herb Tribe. Vol. i. p. 206.

* * * *Perianth single.*

10. Daphné.   *Corolla* 4-cleft.—A shrub bearing leaves only near the summit ρf the stem, intermixed with drooping tufts of green flowers. Daphne Tribe. Vol. ii. p. 159.

Order II.   DIGYNIA.   2 *Pistils.*

11. Polýgonum.   *Perianth* single, deeply 5-cleft: *seed* 1.—Herbaceous plants with alternate leaves and sheathing stipules. Persicaria Tribe. Vol. ii. p. 151.

12. SCLERANTHUS. *Perianth* single, contracted at the mouth, 4-cleft; *seed* 1.—A humble herbaceous plant, with repeatedly forked stems, linear leaves, and small green flowers. KNOT-GRASS TRIBE. Vol. i. p. 226.

13. CHRYSOSPLENIUM. *Perianth* single, *capsule* 2-beaked, many-seeded.—Herbaceous aquatic plants, with juicy stems and leaves, and small greenish yellow flowers. SAXIFRAGE TRIBE. Vol. i. p. 239.

ORDER III. TRIGYNIA. 3 *Pistils.*

14. POLÝGONUM.—Herbaceous plants with alternate leaves and sheathing stipules. PERSICARIA TRIBE. Vol. ii. p. 151.

ORDER IV. TETRAGYNIA. 4 *Pistils.*

15. PARIS. *Sepals* 8, 4 alternately smaller; *fruit* a berry.—An herbaceous plant, with an erect stem, bearing 4 large leaves and a single, stalked, green flower.— HERB-PARIS TRIBE. Vol. ii. p. 216.

16. ADOXA. *Flowers* in a terminal head.—A delicate little plant, with a few bright green leaves and a round head of green flowers. IVY TRIBE. Vol. i. p. 284.

17. ELÁTINÉ. *Sepals* 4; *petals* 4, coloured; *capsule* 4-celled.—A minute aquatic plant, with rooting stems and opposite leaves. WATER-WORT TRIBE. Vol. i. p. 83.

18. MÆNCHIA. *Sepals* 4; *petals* 4, coloured; *capsule* 1-celled.—A humble herbaceous plant, with opposite linear leaves and solitary flowers. CHICKWEED TRIBE. Vol. i. p. 86.

CLASS IX. ENNEANDRIA. 9 *Stamens.*

ORDER I. HEXAGYNIA. 6 *Pistils.*

1. BÚTOMUS.—An aquatic plant with long, sword-shaped leaves, and tall stalks bearing each an umbel of handsome rose-coloured flowers. FLOWERING RUSH TRIBE. Vol. ii. p. 240.

CLASS X. DECANDRIA. 10 *Stamens.*

ORDER I. MONOGYNIA. 1 *Pistil.*

* *Ovary superior.*

1. MONÓTROPA. *Flowers* in a cluster, the terminal one with 5 *sepals*, the lateral ones with 4.—A parasitic plant, consisting of a juicy stalk without leaves, and a drooping cluster of brownish flowers. BIRD'S-NEST TRIBE. Vol. ii. p. 20.

2. PÝROLA. *Sepals* 5 ; *corolla* of 5 deep lobes or petals.—Somewhat shrubby plants, with unbranched stems, simple, smooth, evergreen leaves, and large, often fragrant flowers. BIRD'S-NEST TRIBE. Vol. ii. p. 20.

3. MENZIESIA.
4. ANDRÓMEDA.
5. ÁRBUTUS.

Calyx deeply 5-cleft ; *corolla* of 1 petal, egg-shaped.—Shrubs with evergreen leaves and drooping flowers. HEATH TRIBE. Vol. ii. p. 11.

* * *Ovary inferior.*

6. VACCÍNIUM. *Corolla* of 1 petal.—Small shrubs with simple, alternate leaves, and bell- or wheel-shaped flowers. CRANBERRY TRIBE. Vol. ii. p. 8.

ORDER II. DIGYNIA. 2 *Pistils.*

* *Perianth single.*

7. POLÝGONUM. *Perianth* deeply 5-cleft ; *seed* 1.— Herbaceous plants with alternate leaves and sheathing stipules. PERSICARIA TRIBE. Vol. ii. p. 151.

8. SCLERANTHUS. *Perianth* contracted at the mouth, 5-cleft ; *seed* 1. — A humble herbaceous plant, with repeatedly forked stems, linear leaves, and small green flowers. KNOT-GRASS TRIBE. Vol. i. p. 226.

9. CHRYSOSPLENIUM. *Capsule* 2-beaked, many-seeded. —Herbaceous aquatic plants, with juicy stems and leaves, and small greenish yellow flowers. SAXIFRAGE TRIBE. Vol. i. p. 239.

* * *Perianth double.*

10. SAXÍFRAGA. *Petals* 5; *capsule* opening with valves or teeth.—Mostly alpine herbaceous plants, with tufted leaves and panicled large flowers. SAXIFRAGE TRIBE. Vol. i. p. 239.

11. DIANTHUS.
12. SAPONARIA.
{ *Petals* 5; *capsule* 1-celled, opening with teeth.—Herbaceous plants, with glaucous, opposite leaves, and terminal handsome flowers. PINK TRIBE. Vol. i. p. 85.

ORDER III. TRIGYNIA. 3 *Pistils.*

13. POLÝGONUM. *Perianth* single; *seed* 1.—Herbaceous plants with alternate leaves and sheathing stipules. PERSICARIA TRIBE. Vol. ii. p. 151.

14. SILÉNÉ.
15. STELLARIA.
16. ARENARIA.
17. CHERLERIA.
{ *Sepals* 5; *petals* 5; *capsule* many-seeded.—Herbaceous plants with opposite leaves and terminal flowers. CHERLERIA is often without petals. PINK and CHICKWEED TRIBES. Vol. i. pp. 85, 86.

ORDER IV. PENTAGYNIA. 5 *Pistils.*

* *Perianth inferior; carpels* 5.

18. SIBBALDIA. *Petals* and *stamens* inserted on the mouth of the calyx.—A humble alpine plant, with ternate leaves and yellowish flowers, of which the stamens and pistils are very variable in number. ROSE TRIBE. Vol. i. p. 170.

19. COTYLÉDON.
20. SEDUM.
{ *Petals* and *stamens* inserted in the bottom of the calyx.—Herbaceous plants, remarkable for their thick, fleshy leaves. STONECROP TRIBE. Vol. i. p. 230.

* *Perianth inferior ; capsule* 1.

21. Oxális. *Sepals* 5 ; *petals* 5 ; *capsule* 5-celled ; *seeds* enclosed each in an elastic case.—Herbaceous plants with ternate leaves and delicate flowers. Wood-sorrel Tribe. Vol. i. p. 131.

22. Siléné. 23. Lychnis. 24. Agrostemma. 25. Spérgula. 26. Stellaria. 27. Cerástium. *Sepals* 5 ; *petals* 5 ; *capsule* 1-celled. — Herbaceous plants with opposite leaves and terminal flowers. Pink and Chickweed Tribes. Vol. i. pp. 85, 86.

** *Perianth superior.*

28. Adoxa. *Flowers* in a terminal head.—A delicate little plant, with a few bright green leaves, and a round head of green flowers. Ivy Tribe. Vol. i. p. 284.

Class XI. DODECANDRIA. 12 *to* 18 *Stamens.*

Order I. MONOGYNIA. 1 *Pistil.*

1. Asarum. *Perianth* single, 3-cleft —An herbaceous plant, with 2 shining leaves and a solitary, drooping, greenish flower. Birth-wort Tribe. Vol. ii. p. 161.

2. Lythrum. *Calyx* with 12 divisions, alternately smaller ; *petals* 6.—Herbaceous plants with 4-cornered stems, opposite leaves, and purple flowers. Loosestrife Tribe. Vol. i. p. 217.

Order II. DIGYNIA. 2 *Pistils.*

3. Agrimonia.—A rough herbaceous plant, with pinnate leaves and spiked, yellow flowers. Rose Tribe. Vol. i. p. 170.

Order III. TRIGYNIA. 3 *Pistils.*

4. Reséda.—Herbaceous or somewhat shrubby plants, with furrowed stems, smooth leaves, and terminal spikes

or clusters of greenish flowers. ROCKET TRIBE. Vol. i. p. 71.

ORDER IV. TETRAGYNIA. 4 *Pistils.*

5. TORMENTILLA.—Herbaceous plants with slender stems, quinate leaves, and bright yellow flowers. ROSE TRIBE. Vol. i. p. 170.

ORDER V. DODECAGYNIA.

6. SEMPERVÍVUM.—A succulent, herbaceous plant, with remarkably thick leaves, and a terminal cyme of dull purple flowers. STONECROP TRIBE. Vol. i. p. 230.

CLASS XII. ICOSANDRIA. 20 *or more Stamens placed on the Calyx.*

All the genera in this Class belong to the Natural Order ROSACEÆ. ROSE TRIBE. Vol. i. p. 170.

CLASS XIII. POLYANDRIA. *Many Stamens inserted in the Receptacle.*

ORDER I. MONOGYNIA. 1 *Pistil.*

1. DELPHÍNIUM. *Sepals* 5, coloured, the upper one spurred.—A slender, herbaceous plant with terminal clusters of irregular, blue flowers. CROWFOOT TRIBE. Vol. i. p. 2.

2. PAPÁVER.
3. MECONOPSIS.
4. GLAUCIUM.
5. CHELIDONIUM.
{ *Sepals* 2; *petals* 4.—Herbaceous plants with showy flowers, and abounding in white or yellow juice, which is narcotic. POPPY TRIBE. Vol. i. p. 26.

6. ACTÆA. *Sepals* 4; *petals* 4.—An herbaceous plant with triangular stems, twice-ternate leaves, and terminal clusters of white flowers.—CROWFOOT TRIBE. Vol. i. p. 2.

xlviii DIDYNAMIA.

7. Heliánthemum. *Sepals 5, the two outer smallest,* or sometimes wanting; *petals* 5.—Small shrubs with oblong leaves, and showy, white or yellow flowers. Rock-rose Tribe. Vol. i. p. 72.

8. Tilia. *Sepals 5, equal; petals 5.*—Lofty trees with heart-shaped leaves, and yellow flowers which grow several together from the middle of an oblong bract. Lime Tribe. Vol. i. p. 114.

9. Nymphæa.
10. Nuphar.

{ *Sepals* and *petals* numerous.— Aquatic plants with floating leaves and large white or yellow flowers. Water-lily Tribe. Vol. i. p. 24.

Order II. PENTAGYNIA. *2—6 Pistils.*

11. Reséda. *Flowers* irregular; *capsule* solitary, open at the top.—Herbaceous or somewhat shrubby plants with furrowed stems, smooth leaves, and terminal spikes or clusters of greenish flowers. Rocket Tribe. Vol. i. p. 71.

All the remaining plants in this Order belong to the Crowfoot Tribe. Vol. i. p. 2.

Order III. POLYGYNIA. *Many Pistils.*

All the plants of this Order belong to the Crowfoot Tribe. Vol. i. p. 2.

Class XIV. DIDYNAMIA. *4 Stamens, 2 longer than the other 2.*

Order I. GYMNOSPERMIA. *Ovaries 4, 1-seeded.*

1. Verbena. *Corolla* slightly irregular; *ovaries* united.—An erect herbaceous plant, with a roughish stem, and terminal slender spikes of small lavender flowers. Verbena Tribe. Vol. ii. p. 120.

All the remaining plants of this order belong to the
LABIATE TRIBE. Vol. ii. p. 91.

ORDER II. ANGIOSPERMIA.   *Ovary single,
many-seeded.*

2. LINNÆA. *Corolla* bell-shaped, regular.—A beautiful northern plant with trailing stems, roundish leaves,
and delicate pink flowers, which hang in pairs from the
summit of a slender stalk. WOODBINE TRIBE. Vol. i.
p. 293.

3. OROBANCHÉ.
4. LATHRÆA.

{ *Flowers* irregular, forming a spike
on a leafless, scaly stalk.—Parasitic
herbaceous plants, remarkable for
being entirely destitute of leaves.
BROOM-RAPE TRIBE. Vol. ii. p. 66.

All the remaining plants of this order belong to the
FIGWORT TRIBE. Vol. ii. p. 70.

CLASS XV. TETRADYNAMIA.   6 *Stamens*, 4 *long
and 2 short.*

All the plants in this class belong to the CRUCIFEROUS
TRIBE. Vol. i. p. 35.

CLASS XVI. MONADELPHIA.   *Stamens united by
their filaments into* 1 *set.*

ORDER I. PENTANDRIA.   5 *Stamens.*

1. ERODIUM.
2. GERANIUM.

{ *Ovary* of 5 carpels, placed round a
long awl-shaped beak.—Herbaceous
plants with lobed or compound leaves
and purple flowers. GERANIUM TRIBE.
Vol. i. p. 123.

3. LINUM. *Capsule* 10-celled.—Herbaceous plants
with slender stems, narrow leaves, and delicate flowers.
FLAX TRIBE. Vol. i. p. 106.

ORDER II. DECANDRIA.   10 *Stamens.*

4. GERANIUM. *Ovary* of 5 carpels, placed round a
long awl-shaped beak.—Herbaceous plants with lobed

or compound leaves and purple flowers. GERANIUM TRIBE. Vol. i. p. 123.

2. OXÁLIS. *Capsule* 5-celled.—Herbaceous plants with ternate leaves and delicate flowers. WOOD-SORREL TRIBE. Vol. i. p. 131.

3. ULEX. 4. GENISTA. 5. ONÓNIS. 6. ANTHYLLIS. *Flowers* papilionaceous.—Shrubs or herbs, well distinguished by the peculiar form of their corolla, and by their 2-valved capsules or legumes. PEA AND BEAN TRIBE. Vol. i. p. 138.

ORDER III. POLYANDRIA. *20 Stamens or more.*

All the plants in this order belong to the MALLOW TRIBE. Vol. i. p. 109.

CLASS XVII. DIADELPHIA. *Stamens combined by their filaments into 2 sets.*

ORDER I. HEXANDRIA. *6 Stamens.*

1. CORYDALIS.
2. FUMARIA.

Sepals 2 ; *petals* 4, unequal, one of them swollen or spurred at the base.—Herbaceous plants, often with long climbing stems and clusters of small flowers. FUMITORY TRIBE. Vol. i. p. 32.

ORDER II. OCTANDRIA. *8 Stamens.*

3. POLYGALA. *Sepals* 5, the two inner larger and petal-like ; *petals* combined with the filaments. MILKWORT TRIBE. Vol. i. p. 80.

ORDER III. DECANDRIA. *10 Stamens.*

All the plants in this order have papilionaceous flowers, with the stamens combined into 2 sets of 9 and 1. PEA AND BEAN TRIBE. Vol. i. p. 138.

CLASS XVIII. POLYADELPHIA. *Stamens combined by their filaments into 3 or more sets.*

All the plants in this class belong to the ST. JOHN'S WORT TRIBE. Vol. i. p. 116.

CLASS XIX. SYNGENESIA. *Stamens united by their anthers into a tube ; Flowers compound.*

All the plants in this class belong to the COMPOUND FLOWERS. Vol. i. p. 317.

CLASS XX. GYNANDRIA. *Stamens growing in the pistil, and united with it.*

ORDER I. MONANDRIA. 1 *Stamen, and*

ORDER II. DIANDRIA. 2 *Stamens.*

All the plants in these two orders belong to the ORCHIDEOUS TRIBE. Vol. ii. p. 194.

ORDER II. HEXANDRIA. 6 *Stamens.*

1. ARISTOLOCHIA. *Perianth* single, tubular, curved. —A climbing plant with heart-shaped leaves, and pale yellow flowers, the tube of which is swollen at the base and contracted near the mouth. BIRTHWORT TRIBE. Vol. ii. p. 161.

CLASS XXI. MONŒCIA. *Stamens and Pistils in separate flowers on the same plant.*

ORDER I. MONANDRIA. 1 *Stamen.*

1. EUPHORBIA. Several *barren flowers* and 1 *fertile* contained in a *bract,* or involucre of 1 leaf; *ovary* 3-lobed, 3-seeded.—Herbaceous plants abounding in acrid, milky juice, branched principally above, and bearing many green flowers. SPURGE TRIBE. Vol. ii. p. 164.

2. CALLÍTRICHÉ. *Ovary* 4-lobed ; *styles* 2.—Small aquatic plants with opposite leaves, inconspicuous axillary flowers without petals, and 4-seeded fruit. MARE'S-TAIL TRIBE.   Vol. i. p. 212.

3. ZANNICHELLIA. *Ovaries* 4 ; *styles* 4.—A submersed aquatic plant with numerous slender stems, linear leaves, and inconspicuous green flowers. 4. ZOSTÉRA. *Ovaries* 4 ; *style* 1, 2-cleft.—A submersed marine plant with long, grassy leaves, some of which are sheathing at the base, and contain barren and fertile flowers without petals, arranged alternately in 2 rows on a common stalk.   POND-WEED TRIBE.   Vol. ii. p. 254.

5. ARUM.   *Stamens* and *pistils* arranged in separate rings round a common central column or *spadix.*—A curious herbaceous plant with large, glossy leaves, from among which rises a sheath containing a solid crimson column bearing a ring of stamens about the middle, and a row of pistils beneath. CUCKOO-PINT FAMILY. Vol. ii. p. 249.

ORDER II.   DIANDRIA.   2 *Stamens.*

6. CALLÍTRICHÉ. *Ovary* 4-lobed ; *styles* 2.—Aquatic plants with opposite leaves, inconspicuous, axillary flowers without petals, and 4-seeded fruit.   MARE'S-TAIL TRIBE.   Vol. i. p. 212.

7. LEMNA.   *Ovary* 1-celled ; *style* 1 ; *stigma* 1.— Minute, stemless, aquatic, floating plants, consisting entirely of a few thread-like roots and simple leaves, embedded in which last flowers are produced, but rarely. DUCKWEED TRIBE.   Vol. ii. p. 253.

8. CAREX.   Inflorescence in terminal spikes formed of chaffy scales or *glumes*, with a flower at the base of each.   SEDGE TRIBE.   Not described in these volumes.

ORDER III.   TRIANDRIA.   3 *Stamens.*

9. TYPHA.   Inflorescence in a large terminal spike, the upper part of which contains *stamens* only, the lower,

pistils only. 10. SPARGANIUM. Inflorescence in globular heads, of which the upper ones contain *stamens* only, the lower, *pistils* only.—Aquatic plants with linear leaves, growing on the margins of rivers and ponds. REED-MACE TRIBE. Vol. ii. p. 246.

* Many of the SEDGE TRIBE belong to this order; their inflorescence is in terminal spikes, formed of chaffy scales or *glumes*, with a flower at the base of each. Not described in these volumes.

ORDER IV. TETRANDRIA. *4 Stamens.*

11. LITTORELLA. *Barren flower: sepals* 4; *petals* 4; *stamens* very long.—*Fertile flower: ovary* 1-seeded; *style* very long.—An herbaceous marsh plant with fleshy, linear leaves, a stalked, barren flower, and sessile, fertile flowers. PLANTAIN TRIBE. Vol. ii. p. 140.

12. ALNUS. *Flowers* in catkins, the *scales* of which are 3-cleft.—A tree with smooth jagged leaves and numerous small catkins. CATKIN-BEARING TRIBE. Vol. ii. p. 178.

13. MYRÍCA. *Flowers* in catkins, the *scales* of which are entire.—A shrub, growing in boggy ground, well distinguished by its resinous, aromatic catkins, which appear before the leaves. CATKIN-BEARING TRIBE. Vol. ii. p. 178.

14. BUXUS. *Sepals* 4; *styles* 3; *capsule* with 3 beaks.—A hard-wooded shrub, or small tree, with evergreen leaves and green axillary flowers. SPURGE TRIBE. Vol. ii. p. 164.

15. PARIETARIA.
16. URTÍCA.

Flowers in spikes or clusters; *calyx* of the *barren flower* of 4 sepals, or 4-cleft; *seed* 1.—Herbaceous plants with hairy or stinging leaves, and small, green flowers, which are remarkable for the elasticity of their stamens. NETTLE TRIBE. Vol. ii. p. 172.

ORDER V. PENTANDRIA. *5 Stamens.*

17. ATRIPLEX.—Herbaceous or somewhat shrubby
plants, with fleshy, often angular and mealy leaves, and
leafy compound spikes of greenish flowers, which, as
well as the foliage, are not unfrequently tinged with
red. GOOSE-FOOT TRIBE. Vol. ii. p. 144.

ORDER VI. POLYANDRIA. *More than 5 Stamens.*

18. MYRIOPHYLLUM. *Sepals* 4; *petals* 4; *stamens* 8;
pistils 4.—Submersed aquatics, with finely-divided leaves
and small whorled or spiked flowers. MARE'S-TAIL
TRIBE. Vol. i. p. 212.
19. CERÁTOPHYLLUM. *Sepals* numerous; *petals* 0;
stamens 18—20; *pistil* 1.— Submersed aquatics with
finely divided, rigid, and whorled leaves. HORNWORT
TRIBE. Vol. i. p. 216.
20. SAGITTARIA. *Sepals* 3; *petals* 3.—An herbaceous aquatic, remarkable for its large arrow-shaped
leaves and delicate lilac flowers. WATER-PLANTAIN
TRIBE. Vol. ii. p. 241.
21. POTERIUM. *Flowers* in a round head, the *upper*
fertile, the *lower* barren.—A slender herbaceous plant,
with pinnate leaves and round heads of greenish flowers,
which are remarkable for their tufted pistils and long
stamens. ROSE TRIBE. Vol. i. p. 170.
22. ARUM. *Stamens* and *pistils* arranged in separate
rings round a common central column or *spadix.*—A
curious herbaceous plant with large glossy leaves, from
among which rises a sheath containing a solid crimson
column, bearing a ring of stamens about the middle,
and a row of pistils beneath. CUCKOO-PINT TRIBE.
Vol. ii. p. 249.
23. QUERCUS. 24. FAGUS. 25. CASTÁNEA. 26. BÉ-
TULA. 27. CARPÍNUS. 28. CÓRYLUS. *Barren flowers*
in a catkin or loose spike.—Trees the fruit of which is

a nut, wholly or in part enclosed in a tough case, with the exception of Bétula, the small nut of which is winged. Catkin-bearing Tribe. Vol. ii. p. 178.

Order VII. MONADELPHIA. *Stamens united in 1 set.*

Pinus.—Trees with resinous wood, and linear, often rigid leaves, and bearing their fruit in *cones*. Fir Tribe. Vol. ii. p. 187.

Class XXII. DIŒCIA. *Stamens and Pistils in separate flowers and on different plants.*

Order I. DIANDRIA. *2 Stamens (sometimes 1).*

1. Salix. *Inflorescence* in catkins, the scales of which contain each a single flower; *stigmas* 2 ; *seeds* cottony. —Trees or shrubs with long, flexible branches and simple leaves. Catkin-bearing Tribe. Vol. ii. p. 178.

Order II. TRIANDRIA. *3 Stamens.*

2. Émpetrum. *Perianth* of several scales ; *filaments* very long ; *stigma* 1, rayed.—A small shrub with heath-like leaves, small flowers, and black berries. Crow-berry Tribe. Vol. ii. p. 163.

3. Ruscus. *Flowers* solitary, from the centre of the leaf; *sepals* 6 ; *petals* 0.—A low shrub with green branches, stout leaves, each of which terminates in a thorn, small green flowers, and scarlet berries. Lily Tribe. Vol. ii. p. 218.

4. Valeriana. *Corolla* of one petal, bearing the stamens ; *seed* 1, crowned by the calyx.—An herbaceous plant, with a simple stem, deeply-cut leaves, and a ter-minal corymb of light-pink flowers. Valerian Tribe. Vol. i. p. 308.

5. Salix. *Inflorescence* in catkins, the scales of which contain each a single flower; *stigmas* 2; *seeds* cottony.—Trees or shrubs with long, flexible branches and simple leaves. Catkin-bearing Tribe. Vol. ii. p. 178.

Order III. TETRANDRIA. *4 Stamens.*

6. Viscum. *Petals* 4, fleshy; *fruit* a 1-seeded berry. A parasitic shrub with green branches, leathery leaves, and conspicuous white berries. Mistletoe Tribe. Vol. i. p. 290.

7. Hippópháë. *Barren flower: perianth* 2-cleft;—*fertile flower: perianth* 4-cleft; *fruit* one seeded, berry-like.—A sea-side, thorny shrub, with silvery leaves and orange-coloured fruit. Oleaster Tribe. Vol. ii. p. 158.

8. Rhamnus. *Calyx* 4-cleft.; *petals* 4.—A thorny shrub, with simple serrated leaves, numerous greenish flowers, and black berries. Buckthorn Tribe. Vol. i. p. 135.

9. Myríca. *Flowers* in catkins.—A shrub, growing in boggy ground, well distinguished by its resinous, aromatic catkins, which appear before the leaves. Catkin-bearing Tribe. Vol. ii. p. 178.

10. Urtíca. *Barren flower: sepals* 4; *petals* 0;—*fertile flower: sepals* 2; *petals* 0; *seed* 1.—An erect, little-branched, herbaceous plant, with opposite, stinging leaves, and small green flowers, which are remarkable for the elasticity of their stamens. Nettle Tribe. Vol. ii. p. 172.

Order IV. PENTANDRIA. *5 Stamens.*

11. Húmulus. *Barren flower:* solitary; *sepals* 5; *petals* 0;—*fertile flowers* in large membranaceous catkins. —A climbing plant with rough stems, lobed leaves, and

handsome clusters of drooping catkins. NETTLE TRIBE. Vol. ii. p. 172.

12. RÍBÉS. *Calyx* 5-cleft ; *sepals* 5, inserted on the calyx ; *fruit* a berry.—An alpine shrub with lobed leaves, erect clusters of green flowers, and red berries.— GOOSEBERRY AND CURRANT TRIBE. Vol. i. p. 237.

13. BRYONIA. *Calyx* 5-toothed ; *sepals* 5 ; *filaments* 3 ; *anthers* 5 ; *fruit* a berry.—A climbing plant with elegantly lobed, downy leaves, greenish flowers, and scarlet berries. GOURD TRIBE. Vol. i. p. 222.

14. SALIX. *Inflorescence* in catkins, the *scales* of which contain each a single flower.—A handsome shrub with long, flexible branches, glossy leaves, and seeds tufted with silky down. CATKIN-BEARING TRIBE. Vol. ii. p. 178.

ORDER V. HEXANDRIA. *6 Stamens.*

15. TAMUS. *Perianth* single, in 6 equal segments ; *fruit* a berry.—A climbing plant with heart-shaped, glossy leaves, small green flowers, and scarlet berries. YAM TRIBE. Vol. ii. p. 215.

16. RUMEX. *Seed* 1, 3-cornered, enclosed in the calyx.—Herbaceous plants with erect stems, acid leaves, and panicles of green flowers tinged with red. PERSI-CARIA TRIBE. Vol. ii. p. 151.

ORDER VI. OCTANDRIA. *8 Stamens.*

17. PÓPULUS. *Inflorescence* in catkins, the *scales* of which are jagged; *stamens* 8—30.—Lofty trees remarkable for having the upper portion of the leaf-stalk flattened vertically. CATKIN-BEARING TRIBE. Vol. ii. p. 178.

18. RHODÍOLA. *Sepals* 4 ; *petals* 4.—A mountain plant, with thick, fleshy leaves and yellow flowers, remarkable for the rose-like scent of its root. STONECROP TRIBE. Vol. i. p. 230.

ORDER VII.   ENNEANDRIA.   *9 Stamens.*

19. MERCURIALIS.   *Sepals* 3 ; *petals* 0 ; *capsule* 2-lobed.—Herbaceous plants with roughish leaves, bearing small green flowers in loose spikes.   SPURGE TRIBE. Vol. ii. p. 164.

ORDER VIII.   DECANDRIA.   10 *Stamens.*

20. LYCHNIS.   *Calyx* 5-cleft; *petals* 5.—An herbaceous plant, with downy, oblong leaves, and large, bright-purple or white flowers.   PINK TRIBE.   Vol. i. p. 85.

ORDER IX.   POLYANDRIA.   *Many Stamens.*

21. STRATIOTES.   *Sepals* 3 ; *petals* 3.—A curious aquatic plant, with sword-shaped, prickly leaves, and large, white flowers, which grow from a 2-leaved sheath. FROG-BIT TRIBE.   Vol. ii. p. 192.

22. PÓPULUS.   *Inflorescence* in a catkin, the *scales* of which are jagged.—Lofty trees, remarkable for having the upper portion of the leaf-stalk flattened vertically. CATKIN-BEARING TRIBE.   Vol. ii. p. 178.

ORDER X.   MONADELPHIA.

Stamens united by their filaments into 1 *set.*

23. JUNÍPERUS.   *Berry* formed of the united scales of the catkin, 3-seeded.—A shrub with prickly, evergreen leaves, inconspicuous flowers, and purple berries.   FIR TRIBE.   Vol. ii. p. 187.

24. TAXUS.   *Seed* solitary, half invested with a fleshy cup.—A hard-wooded tree, with linear, evergreen leaves, and scarlet seed-vessels.   FIR TRIBE.   Vol. ii. p. 187.

ORDER XI. POLYADELPHIA.

Stamens combined by their filaments into 3 sets.

25. BRYONIA. *Calyx* 5-toothed; *petals* 5; *filaments* 5.*—A climbing plant, with elegantly-lobed downy leaves, greenish flowers, and scarlet berries. GOURD TRIBE. Vol. i. p. 222.

CLASS XXIII. POLYGAMIA. *Flowers of 3 kinds; some bearing Stamens and Pistils; some, Stamens only; and some, Pistils only.*

ORDER I. MONŒCIA.

All three kinds of flowers on the same plant.

1. ÁTRIPLEX. — Herbaceous, or somewhat shrubby plants, with fleshy, often angular and mealy leaves, and leafy, compound spikes of greenish flowers, which, as well as the foliage, are often tinged with red. GOOSE-FOOT TRIBE. Vol. ii. p. 144.

CLASS XXIV. CRYPTOGAMIA.

This Class, by far the most extensive in the LINNÆAN ARRANGEMENT, comprises all plants in which the existence of stamens and pistils cannot be detected. Here are placed the FERNS, MOSSES, LIVERWORTS, LICHENS, FUNGI, and SEA-WEEDS, none of which are described in these volumes.

ERRATA.

Vol. i. p. 91.   Title of cut; *for* SILENE INFLATA (*Bladder Campion*),
read SILENE MARITIMA (*Sea Campion*).

Vol. ii. p. 38, *for* March, *read* Marsh.

NATURAL ARRANGEMENT OF PLANTS.

Class I. DICOTYLEDONOUS PLANTS.

This is the most extensive class in the vegetable kingdom, and derives its name from the fact that the *seeds* are composed of two, or more, lobes, called *cotyledons*, which enclose the *plumule*, or embryo of the future plant. As germination commences, the plumule lengthens downwards into a root, called, in its early stage, a *radicle :* at the same time the upper extremity lengthens into a *stem*, which is composed of *bark, woody fibre, spiral vessels, cellular tissue,* and a central column of *pith.* The stem increases in diameter by deposits beneath the bark, but *outside* the existing fibre. Hence, the plants belonging to this class are called EXOGENOUS (increasing by additions on the outside). In all trees and shrubs of this class the wood is arranged in concentric layers, the hardest part being nearest the pith. The *leaves* are reticulated, or furnished with a net-work of veins. The *flowers* are furnished with stamens and

B

pistils ; 5 or 4, or some multiple of 5 or 4, being the predominating number of the parts of fructification.

SUB-CLASS I.

THALAMIFLORÆ.

Flowers furnished with calyx and corolla ; *petals* distinct, inserted into the receptacle or *thalamus ; stamens* springing from below the base of the ovary.

NATURAL ORDER I.

RANUNCULACEÆ.—THE CROWFOOT TRIBE.

Sepals generally 5 ; *petals* 5 or more, frequently irregular in form ; *stamens* indefinite in number, inserted on the receptacle ; *ovaries* generally numerous ; *fruit* consisting of several one or many-seeded carpels ; in *Actæa,* a berry.—An extensive tribe of plants, inhabiting for the most part the temperate regions of the globe. All the British species are herbaceous, leaves generally much divided, and flowers showy ; sepals and petals often running into one another, and sometimes extended into spurs. Most of them possess acrid and poisonous properties if taken into the stomach, and not a few produce wounds if applied to the skin. Some species were formerly used in medicine, and the extract of Monkshood is still employed to relieve pain in affections of the nerves. The Hellebore was held in high repute among the ancients, as a specific for madness : the garden-flower known by the familiar name of Christmas Rose belongs to this family. *Ranunculus sceleratus* is one of the most generally diffused plants, being as common in America, and on the banks of the Ganges, as it is in our own marshes.

* *Carpels one-seeded.*

1. CLÉMATIS (Traveller's Joy). — *Sepals* 4—6, resembling petals; *corolla* wanting; *carpels* terminated by a long feathery tail. (Name from the Greek, *cléma*, a vine-shoot.)

2. THALICTRUM (Meadow Rue).—*Sepals* 4—5; *corolla* wanting; *carpels* without tails. (Name from the Greek, *thallo*, to flourish.)

3. ANEMÓNE (Wind-flower).—*Sepals* and *petals* alike, 5—15; *involucre* of 3 cut leaves, distant from the flower. (Name from the Greek, *anemos*, the wind, from the exposed place of growth.)

4. ADÓNIS (Pheasant's Eye).—*Sepals* 5; *petals* 5—10, without a nectary at the base; *carpels* without tails. (Name from *Adonis*, a youth who was killed by a wild boar, and whose blood is fabled to have stained flowers.)

5. RANÚNCULUS (Crowfoot, Buttercup, Lesser Celandine, &c.)—*Sepals* 5 (rarely 3); *petals* 5 (rarely numerous), with a nectary at the base. (Name from the Latin, *rana*, a frog, an animal which frequents the kind of places where these plants grow.)

6. MYOSÚRUS (Mouse-tail).—*Sepals* 5, prolonged at the base into a spur; *petals* 5; *carpels* crowded into a lengthened spike. (Name, Greek for a mouse's tail.)

** *Carpels many-seeded.*

7. TROLLIUS (Globe-flower).—*Sepals* about 15, resembling petals; *petals* 5 or more, small, narrow. (Name said to be derived from an old German word, signifying a *globe*.)

8. CALTHA (Marsh Marigold). —*Sepals* and *petals* alike, 5 or more. (Name from the Greek, *cálathus*, a cup, which its flowers resemble.)

9. HELLÉBORUS (Hellebore). *Sepals* 5, petal-like, per-

sistent; *petals* small, tubular; *carpels* 3—10. (Name
from the Greek, *helein*, to injure, and *bóra*, food.)

10. AQUILÉGIA (Columbine). *Sepals* 5, petal-like,
soon falling off; *petals* 5, tubular, gaping upwards, and
terminating below in a curved, horn-shaped spur; *carpels* 5. (Name from the Latin, *aquila*, an eagle, to the
claws of which its nectaries bear a fancied resemblance.)

11. DELPHÍNIUM (Larkspur).—*Sepals* 5, petal-like,
soon falling off, the upper one helmet-shaped, with a
long spur at the base; *petals* 4, the two upper on long
stalks, and concealed in the spurred sepal; *carpels*
3—5. (Name from *delphin*, a dolphin, to which animal
the upper leaf bears a fancied resemblance.)

12. ACONÍTUM (Monk's-hood).—*Sepals* 5, petal-like,
the upper one helmet-shaped but not spurred; *nectaries*,
2, stalked, tubular at the extremity, and concealed
beneath the helmet-shaped sepal; *carpels* 3—5. (Name
of uncertain origin.)

13. ACTÆA (Bane-berry). *Sepals* 4, soon falling off;
petals 4; *fruit* a many-seeded berry. (Name from the
Greek, *acté*, the elder, from the similarity of the leaves of
the two plants.)

14. PÆONIA (Peony).—*Sepals* 5, unequal; *petals*
5—10; *carpels* 2—5, with fleshy stigmas formed of
two plates. (Name from *Pœon*, a Greek physician,
who is said to have cured wounds with it.)

1. CLÉMATIS.

1. *C. Vitalba* (Traveller's Joy).—The only British
species. A common hedge-shrub, where limestone or
chalk enters largely into the composition of the soil,
climbing other shrubs by the help of its twisting leaf
stalks; well distinguished in summer by its numerous
greenish-white flowers, and in autumn and winter rendered yet more conspicuous by its tufts of feathered
seed-vessels. It received its name from " decking and

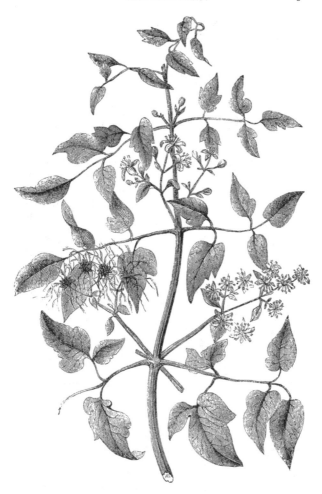

CLEMATIS (*Traveller's Joy*).

adorning waies and hedges where people travel."—
Fl. May, June. Perennial.

2. Thalictrum (Meadow Rue).

1. *T. alpinum*
(Alpine Meadow
Rue). — *Stem* un-
branched; *flowers* in
a simple terminal
cluster, drooping
when fully expand-
ed. — A graceful
little plant, from 4
to 6 inches high,
common on the
mountains of Scot-
land and Wales.—
Fl. June, July. Per-
ennial.

2. *T. minus* (Lesser
Meadow Rue). —
Stem zigzag, branch-
ed; *leaves* thrice
pinnate; *leaflets*
three-cleft, glaucous;
flowers on slender
stalks, drooping.—
In limestone and
chalky pastures, or

THALICTRUM ALPINUM (*Meadow Rue*).

on banks of shell-sand, among bushes; 1—2 feet high.
Often detected, like the following, by its conspicuous
yellow stamens.—Fl. June, July. Perennial.

3. *T. flavum* (Yellow Meadow Rue).—*Stem* erect,
branched; *leaves* twice pinnate; *flowers* crowded, erect.
—Not uncommon about the banks of ditches and rivers;
3—4 feet high.—Fl. June, July. Perennial.

3. A. NEMORÓSA (*Wood Anemone*).

1. *A. nemorósa* (Wood Anemone, Wind-flower).—
Flower drooping; *sepals* or *petals* 6 ; *carpels* without
tails.—This is one of our most beautiful spring flowers,
adorning our upland woods at the season when primroses and violets are in perfection. The petals and

sepals are generally white, but not unfrequently tinged with pink externally; more rarely they are of a delicate sky-blue, both within and without.—Fl. March—May.   Perennial.

2. *A. pulsatilla* (Pasque-flower). — *Flower* slightly drooping; *sepals* or *petals* 6 ; *carpels* with feathery tails. —In high chalky pastures, not nearly so common as the last.   The flowers are of a violet blue, and appear about the season of Easter (*Pâques*), from which the plant derives its name.—Fl. April, May.   Perennial.

* Two other species of Anemone are described by British botanists, *A. apennina* and *A. ranunculoides*, but they are doubtful natives, and rarely met with. The former has blue flowers, of 12 or more *petals* or *sepals*, and the latter has yellow flowers.

4. ADÓNIS (Pheasant's Eye).

1. *A. autumnális.*—The only British species.   A pretty herbaceous plant, with finely cut leaves and bright scarlet flowers, which in shape are very like buttercups.   It occurs as a weed in corn-fields, but is not supposed to be a native.   Fl. Sept., Oct.   Annual.

5. RANUNCULUS (Buttercup, &c.)

* *Flowers white.*

1. *R. aquátilis* (Water Crowfoot).—*Stem* submersed; *lower leaves* deeply cleft into hair-like segments, *upper ones* floating, three-lobed, variously cut.—A very variable plant : when growing in swiftly running water the plant is wholly composed of hair-like leaves ; but when growing in stagnant water it produces flattened leaves as well, and abundance of large showy flowers.—Fl. May—July.   Perennial.

2. *R. hederáceus* (Ivy-leaved Crowfoot).—*Leaves* all rounded and lobed ; *petals* scarcely longer than the calyx ; *stamens* 5—10.—Smaller than the last, growing either in water, or close to the water's edge.—Fl. all the summer.   Perennial.

ADÓNIS (*Pheasant's Eye*).

** *Flowers yellow; leaves undivided.*

3. *R. Lingua* (Great Spear-wort).—*Leaves* narrow, tapering to a point, sessile; *stem* erect.—The largest British species, 2—4 feet high, growing in watery places.—Fl. July. Perennial.

4. *R. Flámmula* (Lesser Spear-wort).—*Leaves* narrow, tapering to a point, slightly stalked; *stem* creeping at the base.—Sides of watery places, much smaller than the last; leaves sometimes clothed with silky hairs.—Fl. June—August. Perennial.

R. FICARIA (*Lesser Celandine*).

5. *R. Ficaria* (Lesser Celandine).—*Leaves* heart-, or kidney-shaped, angular; *sepals* 3; *petals* about 9.— One of our brightest spring flowers, studding every bank with its numerous glossy star-like flowers. It is placed by some botanists in a distinct genus, when it is called *Ficaria verna.*—Fl. March—May. Perennial.

*** *Flowers yellow ; leaves divided ; carpels smooth.*

6. *R. auricomus* (Wood Crowfoot).—*Leaves* smooth, lower ones kidney-shaped, lobed ; upper ones deeply divided ; *petals* unequal in size.—Woods, common. *Flowers* mostly irregular, owing to some of the petals being imperfectly developed.—Fl. April, May. Perennial.

7. *R. sceleratus* (Celery-leaved Crowfoot).—*Leaves* smooth, cut into oblong segments ; *stem* hollow, juicy ; *carpels* collected into an oblong head.—A highly acrid species, from 6 inches to 2 feet high, growing in watery places in most parts of the world. *Leaves* glossy ; *petals* inconspicuous.—Fl. June—August. Annual.

8. *R. bulbosus* (Bulbous Buttercup).—*Calyx* reflexed ; *flower-stalks* channeled ; *root* bulbous.—This is by children confounded with the two following, under the name of Buttercups. Meadows.—Fl. May, June. Perennial.

R. REPENS (*Creeping Buttercup*).

9. *R. repens* (Creeping Buttercup).—*Calyx* spreading ; *flower-stalks* channeled ; *root* creeping. — Moist meadows and sides of ditches. A troublesome weed, increasing by creeping shoots, or *scions*, which take root wherever a leaf is produced.—Fl. June—August. Perennial.

10. *R. acris* (Meadow Crowfoot).—*Calyx* spreading ;
flower-stalks cylindrical, not furrowed ; *root* composed of
long fibres.—Meadows, common.   Well distinguished
from the preceding by the above characters, as well as
its slender stem, and by the narrower segments of its
upper leaves.—Fl. June, July.   Perennial.   A double
variety is common in gardens, under the name of Yellow
Bachelor's Buttons.

**** *Flowers yellow ; leaves divided ; carpels not
smooth.*

11. *R. hirsútus* (Pale Hairy Buttercup).—*Calyx* reflexed ; *root* fibrous ; *carpels* margined, and rough with
small warts.—Meadows and waste ground.   Smaller
than either of the last three, for stunted specimens of
which it might be mistaken.—Fl. June—October.   Annual.   A double variety is sometimes found wild.
12. *R. arvensis* (Corn Crowfoot).—*Calyx* spreading ;
carpels large and prickly.—Cornfields.   Well distinguished by its deeply-cut smooth foliage, as well as by
its prickly carpels.   One of the most poisonous of the
genus.—Fl. June.   Annual.
13. *R. parviflórus* (Small-flowered Crowfoot).—*Stem*
prostrate, hairy; *seeds* covered with small hooked prickles.
—Fields and waste places, not common.   Well distinguished by its hairiness, prostrate mode of growth,
and inconspicuous flowers which grow opposite the
leaves.—Fl. May—August.   Annual.
* Most of the plants in this genus are acrid, and are
said to be injurious to cattle if mixed largely with their
food.   *R. flammula* and *sceleratus* are used in the
Hebrides to raise blisters ; they are, however, of objec-
tionable use, being likely to produce sores difficult to
heal.   *R. aquátilis* is by some botanists separated into
several species.   Another species, *R. alpestris*, which
grows on the Clova mountains, has divided leaves and
white flowers.

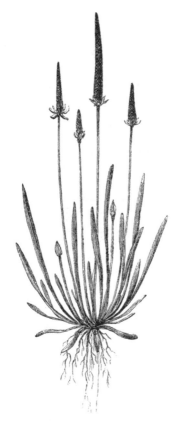

6. Myosúrus (*Mouse-tail*).

1. *M. mínimus* (Common Mouse-tail).—A small annual plant, 3—6 inches high, growing in gravelly or chalky corn-fields in many parts of England, easily distinguished from every other British plant by the arrangement of its ripe carpels into the appearance of a mouse's tail. Fl. May. Annual.

7. Trollius (*Globe Flower*).

1. *Europæus* (Mountain Globe Flower). — A large and handsome plant, common in gardens, and growing wild in the mountainous parts of Scotland, Wales, and the north of England. The flowers are composed of about fifteen concave sepals, which converge into the form of a globe; the petals are shorter and narrower than the sepals, and also curve inwards. By some botanists the sepals are called petals, and the petals nectaries. Fl. June, July. Perennial.

CALTHA (*Marsh Marigold*).

8. CALTHA (Marsh Marigold).

1. *C. palustris* (Common Marsh Marigold).— A showy plant, resembling a gigantic Buttercup, abundant in marshes and by the sides of streams. *Leaves* kidney-shaped, large and glossy. A double variety is common in gardens.

9. HELLÉBORUS (Hellebore).

1. *H. víridis* (Green Hellebore).—*Leaves* digitate ; *sepals* spreading.—A coarse herbaceous plant, remarkable for the light green hue of its flowers. The *petals*, or as some botanists call them, the nectaries, of this plant, as well as of the following, are tubular, shorter than the calyx, and contain honey, which is said, and perhaps with reason, to be poisonous.—Fl. March, April. Perennial.

2. *H. fœtidus* (Stinking Hellebore). *Leaves* pedate ; *sepals* converging.—Best distinguished from the preceding by its evergreen leaves which are not divided to a common centre, and by the purple hue of its sepals. Both species are considered doubtful natives, being generally found in the vicinity of houses.—Fl. March, April. Perennial.

* These two species are remarkable for their green leaf-like petals, and for the large tubular nectaries, in which small flies may sometimes be found caught, as in a trap. *H. niger* is a handsome species, with large white flowers tinged with rose-colour, which are best known by the name of Christmas Rose. This is by some supposed to be the Hellebore of the ancients, so famous as a specific for madness ; but most probably it was that now called *H. officinalis.*

HELLÉBORUS VIRIDIS (*Green Hellebore*).

C

AQUILEGIA (*Columbine*).

10. Aquilegia (*Columbine*).

1. *A. vulgáris* (Common Columbine). — The only British species, common in gardens, to which it is in spring very ornamental, with its delicate, folded leaves ; and no less so in summer, with its curiously shaped flowers, which are of various colours. When growing wild, its flowers are usually blue or white. It may be distinguished from all other British flowers by having each of its five petals terminated in an incurved horn-like spur. It derives its name, Columbine, from the fancied resemblance of its flowers to a nest of doves. Fl. June, July. Perennial.

11. Delphínium (*Larkspur*).

1. *D. Consólida* (Field Larkspur).—A doubtful native, though often found in considerable quantities in sandy or chalky corn-fields. It closely resembles some of the species commonly cultivated in gardens. Fl. June, July. Annual.

12. Aconîtum (*Monk's-hood*).

1. *A. Napellus* (Common Monk's-hood, or Wolf's-bane).—A common garden plant, more remarkable for the curious structure of its tubular nectaries, which are concealed under the hooded upper petal, than for beauty, or any other desirable qualities. Children often amuse themselves by pulling off the hood, and exposing the nectaries, when the remainder of the flower bears a fanciful resemblance to a car drawn by doves. The whole plant (especially the root) is very poisonous, and derives its name of Wolf's-bane from being used, in conjunction with more attractive food, as a bait in wolf-traps. It is scarcely to be considered a native of Britain.—Fl. June, July. Perennial.

ACONÎTUM (*Monk's-hood*).   ACTÆA (*Bane-berry*).

13. Actæa (*Bane-berry*).

1. *A. spicata* (Bane-berry, or Herb Christopher).—The only British species, well distinguished by the

generic characters given above, and by its triangular stem. It is a local plant, having been found only in Yorkshire.—Fl. May. Perennial.

14. PÆONIA (*Peony*).

1. *P. corallina* (Entire-leaved Peony). — Scarcely admissible into the British Flora, as it grows only on an island called the Steep Holmes, in the Severn, into which it was probably introduced. It is one of the herbaceous species cultivated in gardens.—Fl. May, June. Perennial.

ORD. II.—BERBERIDEÆ.—THE BARBERRY TRIBE.

Sepals 3, 4, or 6, in a double row, often coloured, soon falling off, surrounded externally by petal-like scales; *petals* either equal in number to the sepals, and opposite to them, or twice as many, often with a gland at the base; *stamens* equal in number to the petals, and opposite to them; *anthers* opening by a valve from the base upwards; *ovary* solitary, 1-celled, 1—3-seeded,

generally turning to a berry.—Shrubs, growing principally in mountainous parts of the temperate zones,
especially in the north of India. Several species have
thorny stems and astringent bark, and furnish a yellow
dye ; the berries are acid, and may be made into an
agreeable preserve ; those of one species, *Bérberis Asia-
tica,* are dried in the sun like raisins. Several handsome species are cultivated in gardens, under the name
of *Mahonia.*

BÉRBERIS (*Barberry*).

1. Bérberis (Barberry).—*Sepals* 6 ; *petals* 6, with
2 glands at the base of each within ; *berry* 2-seeded.
(Name said to be of Arabic origin.)

2. Epimédium (Barrenwort). —*Sepals* 4 ; *petals* 4,
with a scale at the base of each within ; *pod* many-
seeded. (Name of uncertain origin.)

1. Bérberis (*Barberry*).

1. *B. vulgáris* (common Barberry).—The only British species. A pretty shrub, not uncommon in woods and hedges ; remarkable for the light colour of its bark, which is yellow within, and for its 3-forked spines. The flowers are yellow, and grow in drooping clusters ; the *filaments* are elastic and irritable, so that when touched ever so lightly by the legs of an insect, or by any other small body, they spring forward and close on the pistil ; after some time, they slowly recover their original position. The berries are oblong, red when ripe, and gratefully acid. The shrub is often rooted out by farmers, on account of an erroneous opinion that it is liable to produce rust in wheat.—Fl. June. Shrub.

2. Epimédium (*Barrenwort*).

1. *E. alpínum* (Alpine Barrenwort).—The only species found in Britain ; occurring here and there in mountainous woods in some parts of Scotland, and the north of England, but not considered to be indigenous. Each *stem* bears a single leaf, which is composed of 3 ternate delicate leaflets.—Fl. May. Perennial.

Ord. III.—NYMPHÆACEÆ.—Water Lily Tribe.

Sepals 4—6, gradually passing into *petals,* and these into *stamens,* all being inserted on a fleshy disk, which surrounds the ovary ; *stigma* sessile, rayed ; *berry* many-celled, many-seeded.—Herbaceous aquatic plants, with peltate, floating leaves, and large, often fragrant, flowers. The roots of some species are roasted and eaten : the seeds contain a considerable quantity of starch, and in seasons of scarcity are used as food.   The East Indian *Nelumbium*

NYMPHÆA ALBA (*White Water Lily*).

speciosum is said to have been the sacred bean of Pytha-goras.   Its curious seed-vessels, filled with vegetating seeds, are thought to have originated the cornucopia of the ancients.   One plant of this order, *Victoria regális,* the largest and most beautiful of aquatic plants, pro-duces blossoms 15 inches, and leaves 6 feet in diameter. The seeds are eatable, and are called in South America, Water Maize.

1. NYMPHÆA (Water Lily).—*Sepals* 4 ; *petals* inserted on a fleshy disk. (Name from its growing in places which the nymphs were supposed to haunt.)

2. NUPHAR (Yellow Water Lily).—*Sepals* 5 ; *petals* inserted on the receptacle. (Name of Greek origin.)

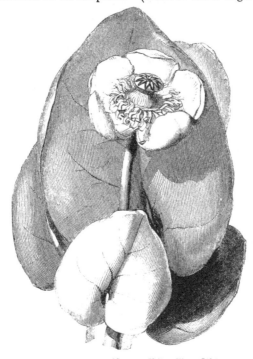

NUFHAR LUTEA (*Common Yellow Water Lily*).

1. NYMPHÆA (*Water Lily*).

1. N. ALBA (White Water Lily).—The only British species, and, perhaps, the most magnificent of our native flowers, inhabiting clear pools and slow rivers. The

flowers rise above the water under the influence of light, and expand only during sunshine, in the middle of the day. Towards evening they close and sink beneath the surface.—Fl. July. Perennial.

2. NUPHAR (*Yellow Water Lily*).

1. *N. lútea* (Common Yellow Water Lily).—*Stigma* with 14—20 rays, which do not extend to the margin.— Rivers and ditches, frequent. Much smaller than the last in all its parts. *Flower* yellow, and nearly globose, smelling like brandy, whence, in Norfolk, and other parts of England, it is called Brandy-bottle. The Turks prepare a cooling drink from the flowers, which they call *Pufer*, (a corruption of the ancient name *Nouphar*).— Fl. July. Perennial.

2. *N. púmila* (Least Yellow Water Lily).—*Stigma* of 8—10 rays, which extend beyond the margin.— Much smaller than the preceding, from which it differs principally in the toothed edge of the stigma. It grows in several of the small Highland lakes.—Fl. July, August. Perennial.

ORD. IV.—PAPAVERACEÆ.—THE POPPY TRIBE.

Sepals 2, soon falling off; *petals* 4 ; *ovary* 1 ; *stigma* rayed, or lobed ; *capsule* 1-celled, many-seeded ; *seeds*

Ovary of the Poppy.

inserted on incomplete partitions, which radiate from the sides of the capsule, but do not meet at the centre.— Herbaceous plants, abounding in a milky juice, which is narcotic, and, under the names of Opium, Laudanum, and Morphia, ranks among the most valuable of medicines. That produced by *Papaver somniferum* is alone used. The seeds of all contain a considerable quantity of oil, which is mild and wholesome.

1. PAPÁVER (Poppy).—*Stigma* sessile, rayed ; *capsule* opening by pores beneath the stigma. (" Named, because it is administered with *pap* (*papa* in Celtic) to induce sleep."—*Sir W. J. Hooker.*)

2. MECONOPSIS (Welsh Poppy).—*Style* short ; *stigma* of few rays ; *capsule* opening by pores beneath the top. (Name in Greek signifying, *bearing resemblance to a Poppy.*)

3. GLAUCIUM (Horn Poppy).—*Stigma* 2-lobed ; *capsule* pod-like, 2-celled, 2-valved. (Name from the *glaucous* hue of the foliage.)

4. CHELIDÓNIUM (Celandine).—*Stigma* 2-lobed ; *capsule* pod-like, 1-celled, 2-valved ; *seeds* crested. (Named from *chelídon;* a swallow, because, as Pliny tells us, that bird discovered that its juice was efficacious in restoring sight to its young when blinded.)

1. PAPÁVER (*Poppy*).
* *Capsules bristly.*

1. *P. Argemóne* (Long Rough-headed Poppy).—*Capsule* club-shaped ; *bristles* erect ; *leaves* twice pinnatifid. —A small species, with light scarlet petals, black at the base, occurring sparingly in corn-fields. (The name Argemóne, from *argos*, slothful, was formerly given to Poppies, from their narcotic effects.)—Fl. June, July. Annual.

2. *P. hýbridum* (Round Rough-headed Poppy).— *Capsule* nearly globular ; *bristles* spreading ; *leaves* twice pinnatifid.—In corn-fields, but not common. *Flowers* deep scarlet.—Fl. June, July. Annual.

** *Capsules smooth.*

3. *P. dubium* (Long Smooth-headed Poppy).—*Capsule* oblong ; *bristles* on the flower-stalks close pressed ;

leaves twice pinnatifid.—In cultivated fields. *Flowers* scarlet.—Fl. June, July.   Annual.

4. *P. Rhœas* (Common Red Poppy).—*Capsules* nearly globular ; *bristles* spreading ; *leaves* pinnatifid, cut.— The common Poppy of corn-fields. *Flowers* scarlet, often black at the base.—Fl. June, July.   Annual.

PAPAVER RHÆAS (*Common Red Poppy*).

5. *P. somniferum* (Opium Poppy).—*Capsule* nearly globular ; *whole plant* smooth and glaucous.—Common in gardens, and sometimes found apparently wild in waste ground ; but its native country is unknown. Opium is procured by puncturing the unripe capsules

of this plant, and collecting the juice, which exudes and hardens. The seeds are destitute of narcotic properties, and afford a wholesome oil, which is said to be much used in adulterating olive oil. *Flowers* usually white, with a purple stain at the base of the petals ; but the colours of garden varieties are endless.—Fl. July, August. Annual.

2. MECONOPSIS (*Welsh Poppy*).

1. *M. Cámbrica* (Yellow Welsh Poppy).—The only British species, easily distinguished from any of the fore-

going by its golden yellow flowers, and juice of the same colour; and from the Horned Poppy, by its slender growth, and green, not glaucous foliage.—Rocky places in Wales, Devonshire, Westmoreland, &c.—Fl. June, July. Perennial.

3. GLAUCIUM (*Horned Poppy*).

1. *G. lúteum* (Yellow Horned Poppy).—*Pod* roughish; *leaves* embracing the stem, wavy, very rough and glaucous.—A handsome plant, conspicuous on the sandy seashore with its hoary foliage, and large yellow flowers. The *pods* are cylindrical, 6—10 inches long, and might at first sight be mistaken for flower-stems bare of leaves; *juice* yellow.—Fl. June—August. Biennial.

* Two other species have been found in England, *M. phœniceum*, and *M. violaceum*; but they are not considered to be indigenous.

4. CHELIDÓNIUM (*Celandine*).

1. *C. majus* (Common or Greater Celandine).—The only British species; not uncommon in waste places, or

among ruins, bearing its yellow flowers, which are much
smaller than those of any others of the Poppy tribe, in

CHELIDÓNIUM MAJUS (*Common Celandine*).

stalked *umbels;* the *leaves* are irregularly pinnate, slightly
hairy, and abound, as well as the rest of the plant, in an

orange-coloured juice, which is a violent acrid poison. It is a popular remedy for warts, and has been employed successfully in removing films from the cornea of the eye ; a property which, Pliny tells us, was discovered by swallows ; and hence it derived its name from *chelídon*, a swallow. According to the same author, it comes into flower at the time when those birds arrive, and fades at their departure.—Perennial.

* The Lesser Celandine is a species of Ranunculus, and bears little resemblance, either in appearance or properties, to the present plant.

Ord. V.—FUMARIACEÆ.—The Fumitory Tribe.

Sepals 2, deciduous ; *petals* 4, irregular, and more or less united and swollen, or spurred at the base ; *stamens* 6, in two sets ; *ovary* 1-celled ; *style* thread-like ; *stigma* lobed ; *seed-vessel* 1 or 2-seeded ; *seeds* shining, crested.— Herbaceous plants, with brittle stems, and watery juice, growing mostly in temperate climates. Closely allied to the Poppies, from which they may well be distinguished by their irregular corollas, and watery (not milky) juice.

1. Corýdalis.—*Petals* 4, of which one is spurred at the base ; *seed-vessel* many-seeded. (Name, the Greek name of *Fumitory*.)

2. Fumaria (Fumitory).—*Petals* 4, of which one is swollen at the base ; *seed-vessel* 1-seeded. (Name from *fumus*, smoke ; the smoke of this plant being said by the ancient exorcists to have the power of expelling evil spirits.)

1. Corýdalis.

1. *C. claviculata* (Climbing Corydalis).—*Stem* climbing ; *leaves* pinnate, ending in branched tendrils.— Bushy places, in many parts of Great Britain. A long and slender plant, with delicate green stems and foliage, rising to the height of several feet, by the help of the

bushes among which it grows. *Flowers* in small clusters, yellowish white.—Fl. June—August. Annual.

* Two other species are naturalized in Britain ; *C. solida*, distinguished by its unbranched stem and purple flowers,

CORYDALIS CLAVICULATA (*Climbing Corydalis*).

and *C. lutea*, not uncommon on old walls ; it is, like the last, destitute of tendrils, and bears bright yellow flowers.

2. FUMARIA (*Fumitory*).

1. *F. capreolata* (Ramping Fumitory).—*Sepals* as broad as the corolla, and half as long ; *fruit* globose, notched. — Hedges and corn-fields, common. Plant

D

generally climbing by the help of its twisted leaf-stalks.
Foliage of a delicate green ; *flowers* pale pink, or
cream-coloured, tipped with purple.—Fl. May—August.
Annual.

FUMARIA OFFICINALIS (*Common Fumitory*).

2. *F. officinalis* (Common Fumitory).—*Sepals* narrower than the corolla ; *fruit* nearly globose, terminating
abruptly.—In fields and waste places, common.   Distinguished from the last by its smaller sepals and petals,
which are rose-coloured, tipped with purple ; it generally grows erect.— Fl. nearly all the year round.
Annual.

* Several smaller varieties of Fumitory are not un-frequently met with, which some botanists consider distinct species, and name as such. In these the fruit is more or less pointed ; and there are other minute differences which cannot be detected without accurate examination. They are described by Hooker, under the name of *F. parviflora* (small-flowered Fumitory).

Ord. VI.—CRUCIFERÆ.—The Cruciferous Tribe.

A very large and important Order, well described by the name *cruciferous*, or cruciform, there being inva-riably 4 *petals*, which are placed cross-wise ; *stamens* 6,

CRUCIFORM FLOWER.

of which two opposite ones are shorter than the rest ; *seed-vessel* either a long pod or *silique*, composed of two valves and a central partition, or a shorter pod, called a *silicle*, or pouch, which is for the most part, but not always, similarly constructed. At the base of the sta-mens are generally two green glands, which secrete honey. Most of the plants in this Order possess, in their wild state, stimulant properties, and an acrid flavour ; in medicine they afford a valuable remedy for

the scurvy. Under cultivation, many of them assume a
succulent habit of growth, and hold the first rank among
esculent vegetables. The various kinds of cabbage,
kale, brocoli, turnip, radish, and cress, are the most
remarkable examples. They contain a great deal of
nitrogen gas, to the presence of which is to be attributed
their unpleasant odour when rotting. Some contain a
large portion of sulphur. Oil is contained in the seeds
of many, in such quantities as to be a valuable article
of commerce. There are about 800 species, one-eighth
of which only are found in America; the remainder, for
the most part, inhabit the cold and temperate regions of
Europe and Asia. Upwards of 200 grow in the frigid
zone, where they form a large proportion of the vegetation.

This Order contains all the plants which were placed
by Linnæus in the Class Tetradynamia; that is, all such
as are distinguished by having 6 stamens, 4 long and
2 short. Modern botanists found the main distinctions
of the genera on the position of the radicle or embryo
root with relation to the cotyledons, or seed-lobes; but
as this arrangement presents difficulties to the young
student in botany, it is not considered advisable to adopt
it here.

* *Seed-vessel, a pouch* (silicle) *or short pod.*

† *Pouch 2-valved, with a central vertical partition.*

1. THLASPI (Penny Cress). — *Pouch* rounded, flat,
notched; *valves* boat-shaped, winged at the back; *seeds*
numerous. (Name from the Greek, *thlao*, to flatten.)

2. CAPSELLA (Shepherd's Purse).—*Pouch* inversely
heart-shaped, flat; *valves* boat-shaped, keeled, but not
winged; *seeds* numerous. (Name, a little *capsa*, or
seed-case.)

3. HUTCHINSIA.—*Pouch* elliptical, entire; *valves* boat-
shaped, keeled, not winged; *cells* 2-seeded. (Named in

honour of *Miss Hutchins*, of Bantry, Ireland, an eminent botanist.)

4. TEESDALIA. — *Pouch* roundish, notched ; *valves* boat-shaped, keeled ; *cells* 2-seeded ; *stamens* having a little scale at the base of each, within. (Named in honour of *Mr. Teesdale*, a Yorkshire botanist.)

5. LEPÍDIUM (Pepper-wort).—*Pouch* roundish ; *valves* keeled ; *cells* 1-seeded ; *petals* equal. (Name from the Greek *lepis*, a scale, from the shape of the pouches.)

6. COCHLEARIA (Scurvy Grass).—*Pouch* globose, or nearly so ; *valves* not flattened ; *seeds* numerous. (Name from *cochlear*, a spoon, from the shape of the leaves.)

7. SUBULARIA (Awl-wort). — *Pouch* oval ; *valves* flattened, boat-shaped ; *seeds* numerous. (Name from *súbula*, an awl, from the shape of the leaves.)

8. DRABA (Whitlow Grass).—*Pouch* oval, or oblong ; *valves* slightly convex ; *seeds* many, in two rows. (Name from the Greek *drabé*, acrid.)

†† *Pouch without a central vertical partition.*

9. CAKÍLÉ (Sea Rocket).—*Pouch* angular, with a horizontal joint ; *lower division* containing a pendent seed, the *upper* an erect seed, soon falling off. (Name of Arabic origin.)

10. CRAMBÉ (Sea Kale).—*Pouch* 2-jointed ; *upper cell* containing one pendent seed, which is supported on a stalk springing from the base of the cell ; *lower joint* seedless. (Name from the Greek *crambé*, cabbage.)

11. CORÓNOPUS (Wart Cress).—*Pouch* 2-lobed, rough, not bursting ; *cells* 1-seeded. (Name from the Greek *coroné*, a crow, and *pous*, a foot, from the shape of the leaves.)

** *Seed-vessel, a silique or long pod.*

† *Pod opening by two valves.*

12. DENTARIA (Coral-root).—*Pod* narrow, pointed ; *valves* flat, nerveless ; *seeds* in a single row, on broad

stalks. (Name, *dens*,   ooth, from the teeth-like divisions of the root.)

13. CARDAMÍNE (Bitter Cress).—*Pod* linear ; *valves* flat, nerveless, separating with an elastic spring ; *seeds* in a single row, on thread-like stalks. (Name from the Greek *cardia*, the heart, and *damao*, to fortify, from its supposed strengthening properties.)

14. ÁRABIS (Rock Cress).—*Pod* linear ; *valves* flat, nerved, or veined ; *stigma* nearly sessile, obtuse ; *seeds* in a single row. (Name from being originally an *Arabian* genus.)

15. TURRÍTIS (Tower Mustard).—*Pod* linear, 2-edged; *valves* prominently nerved ; *seeds* in two rows. (Name from *turris*, a tower, either from its towering form, or its usual place of growth.)

16. BARBARÉA (Winter Cress).—*Pod* linear, 4-angled ; *valves* with prominent nerve ; *seeds* in a single row ; *calyx* erect. (Name from *St. Barbara*, to whom it was anciently dedicated.)

17. NASTURTIUM (Cress).—*Pod* nearly cylindrical, short; *valves* convex, nerveless ; *seeds* irregularly placed in two rows ; *calyx* spreading. (Name from *nasus tortus*, a distorted nose, on account of the pungent properties of the plant.)

18. SISYMBRIUM (Hedge Mustard).—*Pod* rounded, or angular ; *valves* convex, with three nerves ; *stigma* entire ; *seeds* in a single row. (Name, the Greek name of the plant.)

19. ERÝSIMUM (Treacle - Mustard). — *Pod* 4-sided ; *valves* keeled ; *stigma* obtuse, entire, or notched ; *seeds* in a single row, not margined. (Name from the Greek *eruo*, to cure, on account of the supposed virtues of the plant.)

20. CHEÍRANTHUS (Wall Flower).—*Pod* flattened ; *valves* with a prominent nerve ; *stigma* of two spreading lobes ; *calyx* erect, two opposite sepals bulging at the base. (Name of Arabic origin.)

21. MATTHÍOLA (Stock).—*Pod* cylindrical, or flattened ; *stigma* of two converging lobes ; *seeds* generally

with a membranous border. (Name in honour of *Dr. Matthiolus*, an Italian botanist.)

22. BRÁSSICA (Cabbage). — *Pod* nearly cylindrical, beaked ; *valves* 1-nerved ; *seeds* globose, in a single row; *calyx* erect. (Name from the Celtic *Bresic*, a cabbage.)

23. SINÁPIS (Mustard).—*Pod* nearly cylindrical, tapering ; *valves* with 3 or 5 nerves ; *seeds* globose, generally in a single row ; *calyx* spreading. (Name from the Greek *sinápi*, mustard.)

†† *Pod without valves.*

24. RÁPHANUS (Radish).—*Pod* swollen, imperfectly jointed, tapering ; *seeds* globular ; *calyx* spreading. (Name, in Greek, denoting early appearance or quick growth.)

1. THLASPI (*Penny Cress*).

1. *T. arvense* (Mithridate Mustard, or Penny Cress). —*Pouch* round, flat, with very broad wings, and a deep notch ; *seeds* striated ; *leaves* oblong, arrow-shaped at the base, toothed, smooth.—In cultivated or waste ground, but not common. Penny Cress derives its name from the size and shape of its seed-vessels, which resemble silver pennies ; its longer name it received from having been " formerly used in the Mithridate confection, an elaborate hodge-podge, now laid aside."—*Sir J. E. Smith.* The *flowers* are white, and very small in comparison with the pouches.— Fl. all the summer. Annual.

2. *T. perfoliatum* (Perfoliate Penny Cress).—*Pouch* inversely heart-shaped ; *style* shorter than the notch of the pouch ; *seeds* 3 to 4 in a cell, smooth ; *stem-leaves* oblong, heart-shaped at the base.—Limestone pastures in Oxfordshire and Gloucestershire, but rare. *Flowers* white.—Fl. April, May. Annual.

3. *T. alpestre* (Alpine Penny Cress). — *Pouch* inversely heart-shaped, abrupt ; *style* longer than the

notch of the pouch ; *seeds* numerous ; *stem-leaves* arrow-shaped at the base ; *stem* simple.—Mountainous lime-stone pastures in the north of England ; rare.—Fl. June, July. Perennial.

THLASPI ARVENSE (*Penny Cress*).

2. CAPSELLA (*Shepherd's Purse*).

1. *C. Bursa Pastoris* (Common Shepherd's Purse).— The only species. A common weed, to be found in

almost every part of the world, varying considerably in luxuriance, according to soil and situation. In stony ground it grows only a few inches high, but in rich soil as much as two feet. It was known to the ancients by the name of *Thlaspi*, but has recently been separated

CAPSELLA BURSA PASTORIS (*Common Shepherd's Purse*).

from that genus, on account of its wanting the characteristic winged valves. The whole plant is more or less rough with hairs; the root-leaves are pinnatifid, those on the stem oblong, toothed, and arrow-shaped at the base.—Fl. nearly the whole year round. Annual.

3. HUTCHINSIA.

1. *H. petræa* (Rock Hutchinsia).—The only British species. A pretty little plant, from 2 to 4 inches high, growing on limestone rocks in several parts of England

HUTCHINSIA PETRÆA (*Rock Hutchinsia*).

and Wales. The *leaves* are elegantly pinnate; the *petals* white, and scarcely longer than the calyx; the *seeds* two in each cell.—Fl. March, April. Annual.

4. TEESDALIA.

1. *T. nudicaulis* (Naked-stalked Teesdalia).—The only British species. A minute and not inelegant plant, bearing several stems, which terminate in small

corymbs of white *flowers*, with unequal petals, the central *stem* being always bare of leaves. The *leaves* are pinnatifid, and grow in a horizontal direction, closely pressed to the ground.—Dry banks ; not common.—Fl. May. Annual.

TEESDALIA NUDICAULIS (*Naked-stalked Teesdalia*).

5. LEPÍDIUM (*Pepper-wort*).

1. *L. latifolium* (Broad-leaved Pepper-wort).—*Leaves* egg-shaped, pointed, simple, smooth ; *pouch* oval, entire. —In salt marshes, and on the sea-coast ; rare. The largest British species, remarkable for its dull glaucous hue. *Flowers* numerous, small, white, in leafy clusters. —Fl. July. Perennial.

2. *L. ruderále* (Narrow-leaved Pepper-wort).—*Leaves* smooth, *lower ones* pinnatifid, toothed ; *upper ones* linear, entire ; *petals* wanting ; *stamens* 2. — Waste places near the sea : smaller than the preceding.—Fl. June. Annual.

3. *L. campestre* (Field Pepper-wort).—*Leaves* downy, *upper ones* arrow-shaped at the base ; *pouch* rough, with

minute scales; *style* scarcely longer than the notch.—
Fl. July, August.   Annual.

LEPIDIUM LATIFOLIUM (*Broad-leaved Pepper-wort*).

4. *L. Smithii* (Hairy Pepper-wort).—*Leaves* downy,
upper ones arrow-shaped at the base; *pouch* not scaly;
style much longer than the notch.—These last two are

common hedge-plants, of erect growth, and downy habit, made more conspicuous by their hoary foliage and numerous pouches, than by their minute white flowers. *L. campestre* is an annual, and sends up a single stem ; *L. Smithii* is perennial, and sends up several stems, which are woody near the base. The latter is the less common of the two.—Fl. June—August. Perennial.

LEPIDIUM CAMPESTRE (*Field Pepper-wort*).

6. COCHLEÁRIA (*Scurvy Grass*).

1. *C. officinalis* (Common Scurvy Grass).—*Pouch* nearly globose ; *root-leaves* between heart- and kidney-shaped, stalked; *stem-leaves* oblong, sessile, slightly lobed, toothed at the base, all fleshy.—On the muddy sea-shore, common. *Stem* often much branched ; *flowers* in rather large corymbs, white. A smaller variety is common on the Highland mountains, which is made by some botanists a distinct species under the name of *C. Grœnlandica* (Greenland Scurvy Grass).—Fl. May. Annual.

2. *C. Anglica* (English Scurvy Grass).—*Pouch* elliptical, veined ; *root-leaves* oblong, entire, stalked ; *stem-*

COCHLEARIA OFFICINALIS (*Common Scurvy Grass*).

leaves oblong, toothed at the base, sessile.—Sea-shores ;
common. Slenderer than the last, with *leaves* more
entire, and larger *pouches.*—Fl. May—Aug. Annual.

3. *C. Dánica* (Danish Scurvy Grass).—*Pouch* ovate, veined ; *leaves* all stalked, lobed and nearly triangular. —Cliffs and hedges near the sea, very common, much smaller than either of the preceding.—Fl. March—June. Annual.

    * The plants of this genus derive their English name from the relief which they afford to persons suffering

SUBULARIA AQUATICA (*Water Awl-wort*).

from scurvy, a disease to which sailors are particularly liable, in consequence of their being debarred from the use of fresh vegetables. Many other plants of the same tribe possess antiscorbutic properties to an equal degree ; but these are particularly available from always growing near the sea. The use of lime-juice in the navy and merchant service has of late years rendered the attacks of this dreadful disease much less frequent than they used to be. Horseradish, *C. Armorácia*, belongs to this family, but is scarcely wild in Britain.

7. Subularia (*Awl-wort*).

1. *S. aquática* (Water Awl-wort).—The only species. —A small plant common on the banks of Alpine lakes. The roots are composed of long white fibres, the leaves all grow from the roots, and are awl-shaped. The *flowers* are small, and are sometimes perfected under water.—Fl. July. Annual.

DRABA VERNA (*Vernal Whitlow Grass*).

8. Draba (*Whitlow Grass*).

1. *D. verna* (Vernal Whitlow Grass).—*Flower-stalk* leafless; *petals* deeply cloven; *leaves* narrow-pointed, somewhat toothed, hairy.—A humble little plant with scanty foliage, and inconspicuous white flowers, but not without interest from its appearing very early in the year; common on walls and dry banks. A variety is found on Ben Lawers with inflated seed vessels.—Fl. February—May. Annual.

2. *D. aizoídes* (Yellow Alpine Whitlow-grass).—
Flower-stalk leafless, smooth ; *petals* notched, twice as
long as the calyx ; *style* much longer than the stamens ;
leaves narrow, pointed, rigid, glossy, keeled and fringed.
—On rocks and walls at Pennard Castle near Swansea,
where it forms dense tufts, conspicuous with bright
yellow flowers.—Fl. March, April. Perennial.

3. *D. incána* (Twisted Whitlow-grass).—*Stem-leaves*
narrow, toothed, hoary; *petals* entire ; *pouch* twisted.—
Mountainous rocks. Remarkable for the down on its
leaves, which is forked in a star-like manner ; stems
from 4 to 6 inches high, bearing white *flowers.*—Fl.
June, July. Biennial.

4. *D. murális* (Speedwell-leaved Whitlow-grass).
Stem leafy, branched; *leaves* rough, egg-shaped, blunt,
toothed, embracing the stem ; *pedicels* spreading horizontally.—Limestone mountains, not common ; growing
from 6 to 12 inches high ; *flowers* white.—Fl. May.
Annual.

* Another species, *D. rupestris* (Rock Whitlow-grass),
grows on the summits of some of the Highland
mountains, but is very rare. It is found in the crevices
of the rocks, and among moss, scarcely reaching two
inches in height, and forming tufts of soft fringed *leaves*
with several *flower-stems* from the same root.

9. CAKÍLÉ (*Sea Rocket*).

1. *C. marítima* (Purple Sea Rocket).— The only
British species. Common on the sandy sea-shore, where
it grows in a bushy manner, with zigzag branched
stems, bearing fleshy, variously cut, glaucous *leaves,* and
corymbs of lilac *flowers.* The *seed-vessels* are of very
curious construction, each containing two *seeds,* of which
the lower is erect, the upper pendent. — Fl. June—
September. Annual.

10. CRAMBÉ (*Sea Kale*).

1. *C. marítima* (Sea Kale).—This is the plant which

E

is so well known in gardens as an esculent vegetable. The part which is eaten is the leaf-stalk blanched by being kept from the action of light.—It is found on

CAKILE MARITIMA (*Purple Sea Rocket*).

various parts of the sea-coast, and differs in no respect from garden specimens as they appear when the forcing is over.—Fl. June. Perennial.

11. Corónopus (*Wart Cress*).

1. *C. Ruéllii* (Wart Cress, or Swine's Cress).—*Pouch* undivided, rough with little sharp points; *style* pro-

minent.—A common roadside weed, with trailing leafy stems and clusters of very small whitish flowers.—Fl. all the Summer. Annual.

CRAMBE MARITIMA (*Sea Kale*).

2. *C. dídyma* (Lesser Wart Cress).—*Pouch.* notched, of two wrinkled lobes; *style* very short.—A common roadside weed in the South and West of England. It differs from the last in having a more slender stem, and more finely cut leaves. It emits a very powerful smell, like that of Pepper-cress, especially when trodden on, or in hot weather; and is particularly nauseous to the taste.—Fl. all the Summer. Annual.

CORONOPUS DIDYMA (*Lesser Wart Cress*).

12. DENTARIA (*Coral-root*).

1. *D. bulbifera* (Bulbiferous Coral-root).—Well distinguished from any other British plant in the order, by
its purple *flowers,* its whitish toothed *roots,* and dark
purple, scaly *bulbs,* which grow in the axils of the
upper leaves, and falling off when mature produce new
plants.—In shady places ; rare.—Fl. April, May.   Perennial.

* The little bulbs produced by this plant, and some
of the Lily Tribe, are to be considered as modified
leaf-buds, from which they scarcely differ, except in
being easily removed without receiving any injury.

DENTARIA BULBIFERA (*Bulbiferous Coral-root*)

13. Cardamíne (*Bitter Cress*).

1. *C. amára* (Large-flowered Bitter-cress).—*Leaves* pinnate, without stipules ; *root leaflets* roundish, those of the stem toothed or angular ; *stem* creeping at the base ; *style* oblique.—By the banks of rivers and canals ; not common. The *flowers* are large and handsome, white, with purple anthers.—Fl. April, May. Perennial.

2. *C. pratensis* (Cuckoo-flower or Ladies' smock).— *Leaves* pinnate, without stipules ; *root leaflets* roundish, slightly angular, those of the *stem* entire ; *style* straight. A common and very pretty meadow plant with large lilac flowers. " They come with the Cuckoo," says Sir J. E. Smith, " whence one of their English as well as Latin names (*Flos cúculi*) ; and they cover the meadows as with linen bleaching, which is supposed to be that of the other. They are associated with pleasant ideas of spring, and join with the white Saxifrage, the Cowslip, Primrose and Hare-bell to compose many a rustic nose-gay." A double variety is sometimes found wild, which is remarkably proliferous, the leaflets producing new plants when they come in contact with the ground, and the flowers, as they wither, sending up a stalked flower-bud from their centres.—Fl. May. Perennial.

3. *C. impatiens* (Narrow-leaved Bitter Cress).—*Leaves* pinnate ; *stipules* fringed.—Moist rocks in some parts of Scotland and the north of England ; rare.—Fl. May, June. Annual.

4. *C. hirsúta* (Hairy Bitter Cress).—*Leaves* pinnate, without stipules ; *leaflets* stalked, toothed ; *pods* erect. —A common weed everywhere, varying in size according to soil and situation, from six to eighteen inches in height. In dry situations it ripens its seeds in March or April, and withers away ; but in damper places continues in flower all the summer. The leaves and young flower-stems afford an agreeable salad. The *flowers* are white, very small, and often imperfect, and are soon overtopped by the lengthening pods, the valves

CARDAMINE PRATENSIS (*Cuckoo-flower, or Ladies' Smock*).

CARDAMINE HIRSUTA
(*Hairy Bitter Cress*).

ARABIS THALIANA
(*Thale Cress*).

of which, when ripe, curl up with an elastic spring if touched, and fly off, scattering the seeds to a considerable distance.—Fl. all the Summer. Annual.

14. Árabis (*Rock Cress*).

1. *A. thaliána* (Thale Cress, or Wall Cress).—*Leaves* oblong, somewhat toothed, hairy ; *root-leaves* slightly stalked ; *stem* branched ; *pods* angular, twice as long as their stalks.—Common on dry banks and walls. A slender plant, with few *stem-leaves* and minute white *flowers*, from 6 to 12 inches high.—Fl. May—July. Annual.

2. *A. hirsúta* (Hairy Rock Cress).—*Leaves* all rough with hairs, those of the stem numerous, heart-shaped, embracing the stem.—A stiff, erect plant, frequent in many parts of Great Britain on walls and banks. *Flowers* small, white.—Fl. June, July. Biennial.

3. *A. petræa* (Alpine Rock Cress).—*Leaves* smooth, pinnatifid, with smaller lobes at the base; those of the stem simple, sessile.—On rocks in Scotland and Wales. *Flowers* larger than the last, tinged with purple.—Fl. July, August. Perennial.

* Three other species are found in various parts of Britain : *A. stricta* (Bristol Rock Cress), which grows on St. Vincent's Rocks, Clifton, and resembles *A. thaliána*, but has larger *flowers*, and is perennial ; *A. ciliata*, which grows at Connemara, Ireland, and Glen Esk, Scotland, and has *leaves* smooth on both sides, and fringed at the edges ; and *A. Turríta*, which grows on old walls in Cambridge and Oxford. The last is a doubtful native ; its *flowers* are furnished with *bracts*, and the *pods* are curved downwards as they ripen.

15. Turrítis (*Tower Mustard*).

1. *T. glabra* (Smooth Tower Mustard).—The only British species, resembling in habit the preceding genus, from which it is distinguished by having the seeds in

TURRITIS GLABRA (*Smooth Tower Mustard*).

flowering hedge-plant, with delicate green leaves and snow-white flowers. The whole plant emits, when bruised, a nauseous scent of garlic, from which it derives its Latin and English names.—Fl. April—June. Ann.

ERYSIMUM ALLIARIA (*Garlic Treacle-Mustard*).

* Another species, *E. orientale*, (Hare's-ear Treacle-Mustard,) with smooth entire *leaves*, and cream-coloured *flowers*, grows on some parts of the coast of Essex, Suffolk, and Sussex.

CHEIRANTHUS CHEIRI (*Wall-flower*).

20. CHEÍRANTHUS (*Wall-flower*).

1. *C. Cheíri* (Wall-flower).—The only British species, flourishing best on the walls of old buildings, and flowering nearly all the summer, although scantily supplied with moisture. Many beautiful varieties are cultivated in gardens, some of which have double flowers. —Perennial.

MATTHIOLA INCANA (*Hoary Shrubby Stock*).

21. MATTHÍOLA (*Stock*).

1. *M. incána* (Hoary Shrubby Stock).—*Stem* shrubby; *leaves* hoary with down, entire; *pods* without glands.—The origin of the garden Stock. On the southern sea-shore of the Isle of Wight, especially about Niton. *Flowers* light purple.—Fl. May, June. Perennial.

F

2. *M. sinuáta* (Great Sea Stock).—*Stem* herbaceous, spreading ; *leaves* oblong, downy, the lower ones imperfectly lobed ; *pods* rough with prickles.—Sandy seacoasts of Wales and Cornwall ; *flowers* dull purple, fragrant by night.—Fl. August. Biennial.

22. BRÁSSICA (*Cabbage*).

1. *B. campestris* (Common wild Navew).—*Root-leaves* lyrate, toothed, roughish ; *stem-leaves* smooth, heartshaped, tapering to a point ; all somewhat glaucous.— Borders of fields ; common. Often confounded with *Cherlock*, from which, however, it may readily be distinguished by the smoothness and glaucous hue of its upper leaves.—Fl. June, July. Annual.

2. *B. olerácea* (Sea Cabbage).—*Root* stem-like, fleshy ; *leaves* perfectly smooth, glaucous, waved, lobed ; *stem-leaves* oblong, obtuse.—The original of all the varieties of garden cabbage, growing on several parts of the seacoast.—*Flowers* lemon-coloured, larger than the preceding.—Fl. May—August. Biennial.

3. *B. Monensis* (Isle of Man Cabbage).—*Leaves* glaucous, pinnatifid ; *stem* nearly leafless ; *pods* four-angled ; beak 1 to 3-seeded. — Sandy sea-shores on the northwestern coast of Britain.—*Flowers* bright lemon coloured, veined with purple. —Fl. June, July. Perennial.

* Two other species of *Brássica* are described by botanists ; *B. Nápus* (Rape or Cole-seed), which is cultivated for the sake of the oil pressed from its seeds, the refuse being used, under the name of oil-cake, for feeding cattle ; and *B. Napa* (the common Turnip) ; neither of which is really wild. Brocoli and Cauliflower are varieties of Cabbage, which have a tendency to produce an unusual abundance of flower-buds. They are cut for the table before the flowers have had time to expand. In the variety known by the name of Brussels sprouts, the unexpanded leaf-buds are eaten.

BRASSICA CAMPESTRIS (*Common wild Navew*).

SINAPIS ARVENSIS (*Wild Mustard, Cherlock*).

23. SINÁPIS (*Mustard*).

1. *S. arvensis* (Wild Mustard, Cherlock).—*Pods* with many angles, rugged, longer than the awl-shaped beak ; *leaves* toothed, rough.—A common weed in corn-fields, and sometimes springing up profusely from ground which has recently been disturbed, though unknown there before.—Fl. all the summer. Annual.

2. *S. alba* (White Mustard).—*Pods* bristly, rugged, spreading, shorter than the flat two-edged beak ; *leaves* pinnatifid. —Waste ground ; *flowers* large, yellow. The

young leaves of this species are used as salad.—Fl. June. Annual.

PODS OF WHITE MUSTARD.

3. *S. nigra* (Common Mustard).—*Pods* quadrangular, smooth, slightly beaked, close pressed to the stalk ; *lower leaves* lyrate ; *upper* linear, pointed, entire, smooth.—Taller than either of the preceding, but bearing smaller flowers. The seeds yield the mustard of our tables.—Fl. June, July. Annual.

* There are two other British species of *Sinápis* which differ from the preceding in having their *seeds* in two rows ; *S. tenuifolia* (Narrow-leaved Wall Mustard), a bushy, slender herb with smooth *stalks* and *leaves*, and linear, slightly beaked *pods;* it grows on walls, principally in the neighbourhood of large towns, Perennial ; and *S. muralis* (Sand Mustard), which somewhat resembles the preceding, but has a hairy *stem*, and is an Annual.

The Mustard-tree of Scripture is supposed, by some authors, to be a species of *Sinápis* closely resembling the British plants of the same genus. But inasmuch as this plant, though more luxuriant than any with which we are acquainted, never attains the dimensions of a tree, it has with some probability been conjectured that the plant in question is the *khardal* of the Arabs, a tree abounding near the sea of Galilee, which bears numerous branches, and small seeds. having the flavour and properties of Mustard.

RAPHANUS RAPHANISTRUM (*Wild Radish*).

24. RÁPHANUS (*Radish*).

1. *R. Raphanistrum* (Wild Radish).—A bristly, or almost prickly plant, with horizontal, lyrate *leaves*, and rather large, straw-coloured *flowers* veined with purple ; well distinguished when in seed by its jointed one-celled *pods*. A variety named *R. maritimus*, which grows on sea-cliffs, has its leaves composed of small and large leaflets arranged alternately, but is not otherwise distinct. In both varieties the flowers are sometimes almost white.

Ord. VII.—RESÉDACEÆ.—The Rocket Tribe.

Sepals 4—6, narrow ; *petals* unequal, ragged or fringed at the back ; *stamens* 10—24, inserted as well as the petals on an irregular disc, which is placed on one side of the flower ; *stigmas* 3, sessile ; *ovary* 3-lobed, 1-celled, many seeded, open at the summit ; *seeds* in three rows.—Herbaceous or somewhat shrubby plants, with alternate leaves and minute stipules, having their flowers in clusters or spikes. Most of the plants of this order inhabit Europe and the nearest parts of Asia and Africa. *Reséda odorata*, Mignonette, is a native of Egypt, and on account of the delicious perfume of its flowers is admitted into every garden.

1. Reséda (Rocket).—*Calyx* many-parted ; *petals* entire, or variously cut, unequal ; *stamens* numerous ; *capsule* 1-celled, opening at the top. (Name from *resédo*, to calm, from the supposed sedative qualities of some species.)

1. Reséda (*Rocket*).

1. *R. Luteola* (Dyer's Rocket, Yellow weed, or Weld.) —*Leaves* narrow undivided ; *calyx* 4-parted.—Waste places, especially on a chalk or limestone soil. An erect herbaceous plant, 2—3 feet high, with long blunt shining leaves, and terminal spike-like clusters of yellowish flowers, with conspicuous stamens, and short, flattened capsules. It is used to dye wool yellow, or, with indigo, green ; the whole plant, when in flower, being boiled for that purpose.—Fl. July. Biennial.

2. *R. Lútea* (Wild Mignonette). — *Leaves* 3-cleft, lower ones pinnatifid ; *calyx* 6-parted ; *petals* 6, very unequal.—On chalky hills and waste places. More bushy than the last, from which it may be well distinguished by the above characters, as well as by the shorter and broader clusters.—Fl. July, August. Biennial.

RESEDA LUTEOLA (*Dyer's Rocket*).

* Another species, *R. fruticulosa* (Shrubby Rocket),
which is not uncommon in gardens, is occasionally found
in waste places, but has little claim to be considered a
native.   It may be distinguished by the glaucous hue of
its pinnate leaves and by its 5-cleft flowers.

ORD. VIII.—CISTINEÆ.—THE ROCK-ROSE TRIBE.

Sepals 5, unequal, the three inner twisted in the
bud; *petals* 5, twisted, when in bud, in a direction con-

trary to the sepals, soon falling off; *stamens* numerous ; *ovary* single, one- or many-celled ; *style* and *stigma*

HELIANTHEMUM VULGARE (*Common Rock-Rose*).

simple ; *capsule* 3- 5- or rarely 10-valved; *seeds* numerous.—Mostly shrubby, but sometimes herbaceous, plants, often with viscid branches ; *leaves* entire ; *flowers* white,

yellow, or red, lasting a very short time.   The plants of
this order are almost confined to the south of Europe
and north of Africa ; the only species which possesses
any remarkable properties is *Cistus Créticus*, which
affords the balsam called *Gum Ladanum*.

1. HELIÁNTHEMUM (Rock-Rose).—*Sepals* 5, the two
outer either smaller or wanting ; *petals* 5 ; *stamens*
numerous ; *capsule* 3-valved.—(Name from the Greek,
helios, the sun, and *anthos*, a flower, because the flowers
expand in the morning.)

1. HELIÁNTHEMUM (*Rock-Rose*).

1. *H. vulgáre* (Common Rock-Rose).—*Stem* shrubby,
prostrate ; *leaves* with fringed stipules, oblong, green
above, hoary beneath ; *calyx* of 5 leaves, the two outer
very small, fringed.—A beautiful little branching shrub,
with clusters of large bright yellow flowers, frequent in
hilly pastures on a chalky or gravelly soil, where its
flowers only expand during sunshine ; the stamens, if
lightly touched, spread out, and lie down on the petals.—
Fl. July, August.   Perennial.

* There are three other British species of *Helianthemum*, which are all of local occurrence, and rare :—*H. ca-
num* (Hoary Rock-Rose) grows on Alpine rocks, in
Wales and the north of England ; the leaves are destitute of stipules, and very hoary beneath, the flowers are
yellow. *H. guttatum* (Spotted Rock-Rose) is a herbaceous
species and an annual ; the flowers are yellow, with a
blood-red spot at the base of each petal.   *H. polifolium*
(White Rock Rose), a small shrubby species with white
flowers, grows on Brent Downs in Somersetshire, and on
several parts of the sea-coast of Devonshire.

ORD. IX.—VIOLACEÆ.—THE VIOLET TRIBE.

Sepals 5 ; *petals* 5, sometimes unequal ; *stamens* 5 ;
anthers lengthened into a flat membrane ; *style* with an

oblique hooded stigma; *ovary* 1-celled; *seeds* numerous,
in three rows.—A beautiful and important tribe of herbaceous plants or shrubs, strongly marked by the above
characters, inhabiting most regions of the world, except
those parts of Asia whicn are within the tropics.   Those
which grow in temperate regions are mostly herbaceous; but in South America, where they are abundant,
most of the species are shrubs.   The roots of some species
are highly valuable in medicine, furnishing Ipecacuanha,
well known for its sudorific and emetic properties.   The
British species also possess medicinal properties, though
they are rarely used.

1. Víola (Violet).—*Sepals* 5, extended at the base ;
petals 5, unequal, the lower one lengthened into a hollow
spur beneath ; *anthers* united into a tube, two lower ones
furnished with spurs, which are enclosed within the spur
of the corolla; *capsule* with three valves. (*Viola* was the
Latin name of some fragrant flower, which was called by
the Greeks *Ion.*)

1. Víola (*Violet*).

* *Leaves and flowers all springing directly from the root.*

1. *V. hirta* (Hairy Violet).—*Leaves* heart-shaped,
rough, as well as their stalks, with hairs ; *bracts* below the
middle of the flower-stalks ; *sepals* obtuse; lateral *petals*
with a hairy central line.—Common in chalk and limestone districts, or near the sea. *Flowers* various shades of
blue, rarely white, scentless.   Best distinguished from
the sweet violet (to which it is nearly allied) by its very
hairy leaves and capsules, by the position of the bracts,
and by the absence of creeping scions.—Fl. April, May.
Perennial.

2. *V. odoráta* (Sweet Violet). —*Leaves* heart-shaped,
slightly downy, especially beneath ; *bracts* above the
middle of the flower-stalks; *sepals* obtuse; lateral *petals*
with a hairy central line ; *scions* creeping.—One of the
most highly prized of all our wild flowers, unrivalled in

fragrance, delicacy and variety of tinting, and doubly welcome from its appearing so early in spring. The flowers are deep purple, lilac, pale rose coloured, or white, and all these tints may sometimes be discovered on the

VIOLA CANINA *and* V. ODORATA.

same bank. The roots possess the medicinal properties of Ipecacuanha, and the flowers are used as a laxative for children. An infusion of the petals is employed as a chemical test, being changed to red by acids, and by alkalies to green. The flowers are said to communicate their flavour to vinegar in which they have been steeped, and it is also said that they are used in the preparation of the Grand Seignor's sherbets.—Fl. March, April. Perennial.

3. *V. palustris* (March Violet).— *Leaves* heart- or kidney-shaped, quite smooth ; *sepals* obtuse ; *spur* very short ; *root* creeping ; *scions* none.—Bogs and marshy ground, common. *Flowers* delicate lilac, with darker veins ; *leaves* light green, often purplish beneath.—Fl. April—June. Perennial.

** *With an evident stem.*

4. *V. canina* (Dog Violet).—*Stem* channeled, leafy, ascending ; *leaves* heart-shaped, acute ; *sepals* acute ; *stipules* long, toothed and fringed ; *bracts* awl-shaped, entire.—Hedges, heaths, and rocky ground; the most common species.—*Flowers* light blue, purple or white, more abundant, and lasting longer than any of the preceding, but less beautiful, and scentless. This species appears to have received its specific name as a reproach for its want of perfume. A variety with very pale blue flowers and narrow leaves, which occurs not unfrequently on heathy ground, is by some botanists considered a distinct species, and named *V. láctea.*—Fl. April—July. Perennial.

5. *V. tricolor* (Pansy or Heartsease).—*Stem* angular, branched ; *leaves* oblong, crenate ; *stipules* deeply cut, terminal lobe broad crenate.—Cultivated fields.—Very different in habit from any of the preceding, and varying considerably in the size and colour of its flowers, which are, however, most frequently light yellow, either alone, or tinged with purple. The cultivated varieties

6. *V. lútea* (Yellow Mountain Violet, or Mountain Pansy). — *Stem* angular, branched principally at the base ; *leaves* oblong, crenate ; *stipules* deeply cut, terminal lobe narrow entire.—Mountain pastures.—Nearly allied to the preceding, and as variable in the size and colour of its flowers.—Fl. June, July. Perennial.

Ord. X.—DRÓSERACEÆ.—Sundews.

Sepals 5, equal; *petals* 5; *stamens* distinct, either equal in number to the petals, or 2, 3, or 4 times as many; *ovary* single; *styles* 3 — 5, often 2-cleft or branched; *capsule* of 1 or 3 cells, and 3 or 5 valves, which bear the seeds at the middle or at the base.— Delicate, herbaceous, marsh plants, often covered with glands; *leaves* alternate, rolled in at the edges before expansion; *flower-stalks* curled when in bud. The leaves of plants belonging to this order are covered with irritable hairs, by which flies and other small insects are entangled and destroyed. The Sundews are acrid, and impart a red dye to the paper in which they are dried. The leaves of *Dionæa* are furnished with a two-lobed appendage, each half of which has three sharp spines in the middle, and is fringed at the edge. When touched by an insect, these two lobes instantaneously close on the ill-fated intruder, and crush it to death. After a short time they open again in readiness for another victim.

1. Drósera (Sundew).—*Sepals* 5; *petals* 5; *stamens* 5; *styles* 3—5, deeply cleft; *capsule* 1-celled, 3—5 valved.—(Name from the Greek *drosys*, dew, the leaves being covered with red hairs, which exude drops of viscid fluid, especially when the sun is shining, and appear as if tipped with dew.)

1. Drósera (*Sundew*).

1. *D. rotundifolia* (Round-leaved Sundew).—*Leaves* all from the root, spreading horizontally, round; *leaf-stalks* hairy; *seeds* chaffy.— An exceedingly curious little plant, 2—6 inches high, growing in bogs. The root is small and fibrous, and takes a very slight hold in the ground; the leaves are densely covered with red hairs, each of which is tipped with a drop of viscid

fluid ; from the centre of the tuft of leaves rises a wiry leafless stalk, bearing several small whitish flowers which only expand in sunny weather ; the flowers are all on

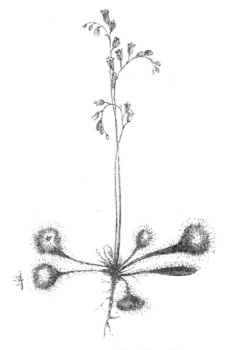

DROSERA ROTUNDIFOLIA (*Round-leaved Sundew*).

one side of the stalk, which in its early stage is curled up, and gradually uncoils itself as the flowers severally expand.—Fl. July, August.   Perennial.

2. *D. longifolia* (Long-leaved Sundew).—*Leaves* all from the root, erect, elongated, broad at the extremity, and tapering towards the base ; *leaf-stalks* smooth ; *seeds*

with a rough, not chaffy coat.—Smaller than the last, and growing with it in boggy places.— Fl. July, August. Perennial.

3. *D. Anglica* (Great Sundew).—*Leaves* all from the roots, erect, oblong, on very long smooth stalks ; *seeds* with a loose chaffy coat.—Stouter and taller than the last, and growing in similar situations, but rare.—Fl. July, August. Perennial.

Ord. XI.—POLYGALEÆ.—The Milkwort Tribe.

Sepals 5, unequal, the two inner larger, generally petal-like ; *petals* 3—5, unequal, more or less combined with the filaments ; *stamens* 8, in two equal sets ; *anthers* 1-celled, opening by pores at the summit ; *pistil* 1 ; *capsule* 1—3 celled ; *seeds* pendulous.—An extensive tribe of herbaceous or shrubby plants, with clustered, often showy flowers. Many are bitter, and their roots are milky. Medicinally they are said to be useful in affections of the lungs, and to excite perspiration. The most celebrated is a North American herb, *Polygala Senega* (Snake-root), to which extraordinary virtues are ascribed. Several species are said to cure snake-bites. *Krameria* (Rhatany-root) is astringent, and furnishes a red infusion used to adulterate Port wine. Some of the above properties, but in a less degree, reside in the only British species, *Polýgala vulgaris.*

1. Polýgala (Milkwort).—*Sepals* 5, the two inner coloured, wing-shaped ; *petals* combined with the filaments, the lower one keeled ; *capsule* flattened, 2-celled, 2-valved ; *seeds* downy, crested at the base. (Name from the Greek, signifying *much milk*, the juice of the root being milky.)

1. Polýgala *(Milkwort).*

1. *P. vulgáris* (Common Milkwort).—Lower *petal* crested in a star-like manner ; *calyx-wings* about equal

in length to the corolla ; *bracts* 3 at the base of each flower ; *stems* simple, ascending, herbaceous ; *leaves* narrow.—Common on heaths and dry pastures, where it is highly ornamental during the later summer months, with its starlike, blue, pink or white flowers.—Fl. June —August. Perennial.

POLYGALA VULGARIS (*Common Milkwort*).

* In the chalk districts a *Polýgala* occurs, whioh is made by some botanists a distinct species, under the rather alarming name of *P. calcárea* (Chalk Milkwort). It differs from the preceding in having the lower leaves largest, and blunt calyx-wings which are differently veined, and in ceasing to flower almost before the common Milkwort begins.

FRANKENIA LÆVIS (*Smooth Sea-Heath*).

Ord. XII.—FRANKENIACEÆ.—Sea-Heaths.

Sepals 4—6 united into a furrowed tube, not falling off; *petals* equal in number to the sepals, furnished with claws, and usually having scales at the junction of the claw and limb; *stamens* equal in number to the petals; *ovary* 1; *style* thread-like, 2, 3, or 4 cleft; *capsule*

1-celled, 2, 3, or 4 valved ; *seeds* very minute, attached to the edges of the .valves.—Herbaceous or somewhat shrubby plants, with branched stems, opposite leaves, which have a membranous sheathing base, and numerous small sessile flowers, inhabiting principally the north of Africa and south of Europe.

1. FRANKÉNIA (Sea-Heath). — *Style* 3-cleft ; *lobes* oblong, with the stigma on their inner side ; *capsule* 3—4 valved. (Name from *John Franken*, a Swedish botanist.)

1. FRANKÉNIA (*Sea-Heath*).

F. lœvis (Smooth Sea-heath).—*Leaves* narrow, rolled back at the edges, smooth, fringed at the base ; *flowers* terminal, or from the forks of the stem.—A small procumbent plant, with wiry stems, crowded leaves, and pale rose - coloured flowers, growing in muddy salt marshes on the eastern coasts of England.—Fl. July. Perennial.

* Another species, *F. pulverulenta* (Powdery Sea-heath), formerly grew on the sea-coast of Sussex, but is now extinct.

ORDER XIII.—ELATINEÆ.—WATER-WORT TRIBE.

Sepals 3—5, distinct, or growing together at the base ; *petals* equal in number to the sepals ; *stamens* equalling, or twice as many as the petals ; *ovary* with 3—5 cells, and as many styles and globular stigmas ; *capsule* with 3—5 cells and valves ; *seeds* wrinkled, springing from the centre of the capsule.—Minute, annual, aquatic herbs, with rooting stems and opposite leaves ; found in most parts of the world.

1. ELÁTINÉ (Water-wort). — *Sepals* 3—4, growing together at the base ; *petals* 3—4 ; *stamens* 3—4, or 6—8 ; *styles* 3—4 celled, many-seeded ; *seeds* cylindrical, furrowed, and transversely striated. (Name of doubtful origin.)

1. ELÁTINÉ (*Water-wort*).

1. *E. hexandra* (Six-stamened Water-wort).—*Flowers* stalked ; *petals* 3 ; *stamens* 6 ; *capsule* 3-celled ; *seeds* straight.—A minute plant, forming turfy beds on the

ELATINE HEXANDRA (*Six-stamened Water-wort*).

margin of lakes, or growing entirely submersed. When left by the subsiding water it assumes a bright red hue, but the flowers are at all times inconspicuous. Rare.— Fl. July—September. Annual.

2. *E. Hydropiper* (Eight-stamened Water-wort).— *Flowers* sessile ; *petals* 4 ; *stamens* 8 ; *capsule* 4-celled ; *seeds* curved. — Yet rarer than the preceding, and growing in similar situations.—Fl. July—September. Annual.

ORDER XIV.—CARYOPHYLLEÆ.

Sepals 5 or 4, distinct, or connected into a tube ; *petals* equal in number to the sepals ; *stamens* usually twice as many as, sometimes equalling, the petals, and like them inserted on the stalk or ring of the ovary ; *ovary* 1, raised on a short stalk, or inserted in a ring ; *stigmas* 2 — 5, running along the inner surface of the styles; *capsule* 1- or imperfectly 2—5-celled, opening by twice as many teeth, or valves, as thére are styles ; *seeds* inserted on a central column.—An extensive and well-marked order of herbaceous plants, inhabiting the temperate and frigid regions of the globe, and not unfrequently bearing ornamental flowers. The stems are swollen at the joints : the leaves always opposite and undivided, and frequently of a glaucous hue. Among garden flowers, the Pink, Carnation, Sweet-William, and Scarlet Lychnis, all belonging to this order, are well known ; and our hedges are much indebted for their showy appearance in spring to the great White Stitch-wort, and in summer to the Red and White Robin. Botanists have distributed the plants of this order into two groups, or sub-orders.

Sub-order I.—Siléneæ.—*Pink Tribe.*

Sepals connected into a tube ; *stamens* united at the base with the stalk of the ovary.

* *Calyx 5-cleft ; petals 5, with long claws ; stamens* 10.

1. Dianthus (Pink).—*Calyx* with 2 or more opposite scales at the base outside ; *styles* 2 ; *capsule* 1-celled, opening at the top with four valves; *seeds* flattened. (Name in Greek signifying *the flower of Jupiter,* from its beauty and fragrance.)

2. Saponaria (Soapwort).—*Calyx* naked at the base ; *styles* 2 ; *capsule* 1-celled, opening at the top with 4 valves ; *seeds* rounded. (Name from *sapo,* soap, the plant abounding in a soapy juice.)

3. Siléné (Catch-fly).—*Calyx* naked at the base ; *petals* generally crowned at the top of the claw ; *styles* 3 ; *capsule* imperfectly 3-celled, opening at the top with 6 valves. (Name of doubtful origin. The English name was given in consequence of flies being often

caught in the viscid fluid which, in some species, surrounds parts of the stem.)

4. Lychnis (Campion).—*Calyx* naked at the base; *petals* generally crowned at the top of the claw; *styles* 5; *capsule* opening at the top with 5 or 10 teeth. (Name from the Greek *lychnos*, a lamp; " the thick cottony substance on the leaves of some species, or some similar plant, having been employed as wicks to lamps." —*Hooker.*)

5. Agrostemma (Corn-cockle).—*Calyx* naked at the base, tough, with leaf-like divisions; *petals* undivided; *capsule* opening at the top with 5 teeth. (Name signifying in Greek, *Crown of the Field.*)

Sub-order II.—Alsíneæ.—*Chickweed Tribe.*

Sepals distinct; *stamens* inserted into a ring beneath the capsule, which is not stalked.

6. Sagína (Pearl-weed).—*Sepals* 4, spreading when in fruit; *petals* 4, sometimes wanting; *stamens* 4; *styles* 4; *capsule* 4-valved. (The name in Latin signifies *fattening meat,* but is totally inapplicable to the minute plants of this genus.)

7. Mænchia.—*Sepals* 4, erect; *petals* 4; *stamens* 4; *styles* 4; *capsule* opening at the top with eight teeth. (Named in honour of *Conrad Mœnch,* Professor of Botany at Hesse Cassel.)

8. Holósteum (Jagged Chickweed).—*Sepals* 5; *petals* 5, toothed at the margin; *stamens* 3—5; *styles* 3; *capsule* opening at the top with 6 teeth. (The name signifies in Greek, *all bone;* but why it was given is uncertain.)

9. Spérgula (Spurrey).—*Sepals* 5; *petals* 5, entire; *stamens* 10 or 5; *styles* 5; *capsule* 5-valved. (" Named from *spargo,* to scatter; from the seeds being so widely dispersed."—*Hooker.*)

10. Stellária (Stitchwort).—*Sepals* 5; *petals* 5, deeply 2-cleft; *stamens* 10; *styles* 3; *capsule* opening

with 6 valves, or teeth. (Name from *stella*, a star, which the expanded flowers resemble in shape.)

11. ARENARIA (Sandwort). — *Sepals* 5 ; *petals* 5, entire ; *stamens* 10 ; *styles* 3 ; *capsule* opening with 6 valves. (Name from the Latin *aréna*, sand, many species growing in sandy ground.)

12. CERASTIUM (Mouse-ear Chickweed).—*Sepals* 5 ; *petals* 5, 2-cleft ; *stamens* 10 or 5 ; *styles* 5 ; *capsule* tubular, opening at the end with 10 or rarely 5 teeth. (Name from the Greek *ceras*, a horn, from the shape of the capsule in some species.)

13. CHERLERIA (Cyphel).—*Sepals* 5 ; *petals* 0, or 5, exceedingly minute, notched ; *stamens* 10, the 5 outer with glands at the base ; *styles* 3 ; *capsule* 3-valved. (Name from *J. H. Cherler*, an eminent botanist.)

1. DIANTHUS (*Pink*).

* *Flowers clustered.*

1. *D. Arméria* (Deptford Pink).—*Stem* and *leaves* downy ; *flowers* in close tufts; *calyx scales* very narrow, downy, as long as the tube.—Waste places; rare. From 1 to 2 feet high, with rose-coloured scentless flowers, dotted with white.—Fl. July, August. Annual.

2. *D. prólifer* (Proliferous Pink). — *Stem* smooth ; *leaves* roughish at the edge ; *flowers* in heads ; *calyx scales* membranous, pellucid.—Gravelly pastures ; rare. Growing about a foot high, and readily distinguished by its heads of rose-coloured flowers, only one of which opens at a time, and by the brown dry scales in which the heads of flowers are enclosed.

** *Flowers not clustered.*

3. *D. caryophýllus* (Clove Pink, Carnation, or Clove Gilly-flower, *that is*, July flower).—*Flowers* solitary ; *calyx scales* 4, broad, pointed, one-fourth as long as the calyx ; *petals* notched ; *leaves* linear, glaucous, with

smooth edges.—On old walls, but scarcely indigenous. Well known, both in the gardens of cottagers and of professed florists, where it is subject to countless varieties,

DIANTHUS ARMERIA (*Deptford Pink*).

all of which are fragrant and beautiful.—Fl. July. Perennial.

4. *D. Cœsius* (Mountain Pink).—*Flowers* mostly solitary; *calyx scales* 4, blunt, one-fourth as long as the calyx; *petals* jagged; *leaves* linear, glaucous, with rough edges.—Limestone Cliffs, Chedder, Somersetshire. *Flowers* rose-coloured, fragrant.—Fl. July. Perennial.

5. *D. deltoídes* (Maiden Pink).—*Flowers* solitary; *calyx scales* usually 2, tapering to a point, half as long as the calyx; *petals* notched; *stem* and *leaves* roughish. —Gravelly banks, but not common. A much branched plant, with stems 6—12 inches high, and rose-coloured flowers with white spots, and a dark ring in the centre; scentless.—Fl. July, August. Perennial.

2. SAPONÁRIA (*Soap-wort*).

1. *S. officinalis* (Common Soapwort).—A robust plant, 2—4 feet high, with broad, pointed, smooth, leaves, and handsome flesh-coloured flowers, which are often double. —It is generally found in the neighbourhood of cultivated ground, and is not considered a native.—Fl. August, September. Perennial.

3. SILÉNÉ (*Catchfly*).

* Stems tufted, short; flowers solitary.

1. *S. acaulis* (Moss Campion).—*Stem* much branched, tufted; *leaves* narrow, fringed at the base; *petals* crowned, slightly notched.—Confined to the summits of the loftiest British mountains, where it forms a densely matted turf, copiously decorated with bright purple flowers.—Fl. June, July. Perennial.

** Stem elongated; flowers panicled; calyx inflated, bladder-like.

2. *S. infláta* (Bladder Campion).—*Stem* erect; *leaves* oblong, tapering; *flowers* panicled, drooping; *calyx* inflated, bladder-like, with a net-work of veins; *petals* deeply cloven, rarely crowned.—A common weed in

corn-fields and pastures, growing from 1 to 2 feet high, and well marked by its numerous white flowers and veined calyces, often tinged with purple. The foliage

SAPONARIA OFFICINALIS (*Common Soap-wort*).

and stem are glaucous, and generally smooth; but a variety which is downy all over is occasionally found.— Fl. June—August. Perennial.

3. *S. maritima* (Sea Campion). — *Stems* numerous from the same root, spreading ; *leaves* oblong, taper-

ing, finely toothed at the margin ; *flowers* few on each stem, or solitary ; *petals* slightly cloven, crowned.— Resembling the last, but of humbler stature, though bearing larger flowers. Common near the sea-shore, as well

SILENE INFLATA (*Bladder Campion*).

as by the sides of mountain rivulets. A variety, with handsome double flowers, has been found in Devonshire. —Fl. all the summer. Perennial.

*** *Stems elongated ; flowers in whorls.*

4. *S. Otites* (Spanish Catchfly).—*Stems* erect, with opposite, tufted branches ; *stamens* and *pistils* on separate plants ; *petals* narrow, entire, not crowned.— Sandy fields in the east of England. The stems are about a foot high, viscid at the middle ; flowers small, yellowish.—Fl. July. Perennial.

**** *Stems elongated; flowers in leafy clusters, alternate.*

5. *S. Anglica* (English Catchfly).—Whole plant hairy
and viscid ; *leaves* narrow, tapering ; *flowers* lateral,
alternate, erect, the lower ones when in fruit reflexed ;
petals crowned, slightly cloven.—Not uncommon in many
parts of England, varying from 6 to 12 inches high or
more, according to soil.   The flowers are inconspicuous,
and of a pinkish white hue.—Fl. all the summer.
Annual.

6. *S. nutans* (Nottingham Catchfly).—*Flowers* all
drooping one way ; *branches* opposite, 3-forked ; *calyx*
much swollen ; *petals* deeply cloven, crowned.—On limestone and chalk rocks; not common.   It grows about
1½ foot high, with large fragrant flowers, which expand
in the evening.—Fl. June, July.   Perennial.

7. *S. cónica* (Striated Corn Catchfly).—*Stem* erect,
forked ; *leaves* narrow, downy : *petals* cloven, crowned ;
calyx of the fruit conical, with 30 furrows.—In sandy
fields, very rare ; from 6 to 12 inches high ; flowers
small, reddish.—Fl. July.   Annual.

8. *S. noctiflóra* (Night-flowering Catchfly).—*Stem*
erect, repeatedly forked ; *calyx* with long teeth, oblong
when in fruit, 10-ribbed.—Sandy and gravelly fields, not
common.   Shorter than the last, and with larger flowers,
which expand about sun-set, and close early in the
morning, and are very fragrant during the night.—Fl.
July.   Annual.

4. LYCHNIS (*Campion*).

1. *L. Flos Cúculi* (Ragged Robin).—*Petals* deeply
4-cleft, crowned ; *leaves* very narrow ; *flowers* loosely
panicled. —A pretty and well-known meadow plant, with
purplish green, angular stems, the lower part of which is
roughish with short bristly hairs, the upper parts lightly
viscid ; flowers rose-coloured, with deeply-cut narrow

segments.—Fl. when the cuckoo is in full song, hence its Latin name *Flos Cúculi.* Perennial.

LYCHNIS FLOS CUCULI (*Ragged Robin*).

2. *L. dioíca* (Red or White Robin, or Campion).— Stameniferous and pistiliferous flowers on different plants ; *petals* 2-cleft half-way down, crowned ; *leaves* oblong, tapering, downy, as well as the stem.—An ornamental hedge plant 2—3 feet high, with rose-coloured or white flowers. The latter variety, which is

made by some botanists a distinct species, under the name of *L. vespertina* (Evening Campion), is fragrant in the evening.—Fl. all the summer. Perennial.

 * Two other species of *Lychnis* are natives of Britain;

AGROSTEMMA (*Corn-Cockle*).

L. viscaria (Red German Catchfly), which is found in Montgomeryshire and in a few places in Scotland, distinguished by its slightly notched petals, 5-celled capsules, and clammy stem : and *R. Alpina* (Red Alpine

Campion), a much smaller species, 5—6 inches high, which grows on the summits of the Clova mountains.

5. AGROSTEMMA (*Corn-Cockle*).

1. *A. Githágo* (Corn-Cockle).—*Calyx* much longer than the corolla ; *petals* undivided, destitute of a crown. —A common corn weed, with an upright downy stem, and large handsome purple flowers ; seeds large, and consequently difficult to be separated from the corn with which they are threshed.—Fl. June, July. Perennial.

6. SAGÍNA (*Pearl-wort*).

1. *S. procumbens* (Procumbent Pearl-wort).—Perennial ; *stems* prostrate, smooth ; *leaves* pointed ; *petals*

SAGINA PROCUMBENS, *and* S. APETALA.

much shorter than the calyx ; *capsules* curved downwards before ripening.—Well known to gardeners as a troublesome weed infesting the paths, and so prolific as to require repeated eradication. The flowers are at all times inconspicuous ; the stems are from 1 to 4 inches long.—Fl. all the summer. Perennial.

* There are two other British species, both very like the preceding : *S. apétala* (Annual Pearl-wort) distin-

guished by its slightly hairy, erect stems, and fringed leaves; the capsules, too, are erect: and *S. maritima* (Sea Pearl-wort), which has blunt fleshy leaves, and flowers destitute of petals.

7. MÆNCHIA.

1. *M. erecta* (Upright Mænchia) —·A small upright plant 2—4 inches high, with rigid glaucous *leaves*, and

MÆNCHIA ERECTA (*Upright Mænchia*).

white *flowers*, which are large in proportion to the rest of the plant; the *sepals* are sharp pointed, with a white membranous edge; the *petals* expand only in the sunshine.—Fl. May. Annual.

HOLOSTEUM UMBELLATUM (*Umbelliferous Jagged Chickweed*).

8. HOLÓSTEUM (*Jagged Chickweed*).

1. *H. umbellatum* (Umbelliferous Jagged Chickweed). —A singular little plant, 4—5 inches high, with leafy *stems* which are smooth below, and hairy and viscid

between the joints above. The *flowers* grow in terminal umbels about 5 together, and are bent back after flowering; *petals* white, with a reddish tinge —Very rare, on old walls at Norwich and elsewhere.—Fl. April. Annual.

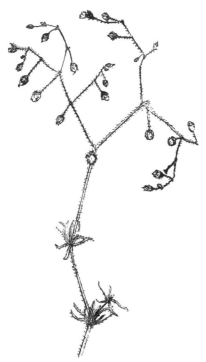

SPERGULA ARVENSIS (*Corn Spurrey*).

9. SPERGULA (*Spurrey*)

1. *S. arvensis* (Corn Spurrey).—*Leaves* cylindrical, in whorls, with minute chaffy stipules at the base; *flowers*

panicled, bent down when in fruit.—A common weed in gravelly corn-fields 6—12 inches high, flowers white.— " Cattle are fond of this plant, and it is an object of culture in Holland." *Sir W. J. Hooker.*—Fl. all the summer. Annual.

2. *S. nodósa* (Knotted Spurrey).—*Leaves* opposite, growing together at the base, upper ones very short, growing in knots ; *flower-stalks* always erect ; *petals* longer than the calyx.—Wet sandy places ; not uncommon. A pretty little plant 3—4 inches high, with conspicuous white flowers, 2 or 3 together, and tufted leaves which at once distinguish it from any other British plant.

* *S. saginóides* (Peart-wort Spurrey) and *S. subulata* (Awl-shaped Spurrey), the former a native of the Scotch mountains, the latter not uncommon in gravelly pastures, have solitary flowers on long flower-stalks, and in habit approach very nearly to *Sagina procumbens.*

10. STELLARIA (*Stitch-wort*).

1. *S. média* (Chickweed).—*Leaves* egg-shaped, with a short point ; *stems* with a hairy line alternating from side to side ; *petals* deeply 2-cleft, not longer than the sepals ; *stamens* 5—10.—Roadsides, waste places, and as a weed in gardens ; abundant. *Leaves* succulent ; *flowers* small, white. Well distinguished by a hairy line which runs up one side of the stem, and when it reaches a pair of leaves, is continued on the opposite side.— Fl. all the year round. Annual.

2. *S. Holostea* (Greater Stitchwort, Satin-flower, or Adder's meat).—*Stem* nearly erect, angular, rough-edged ; *leaves* very narrow, tapering to a long point, minutely fringed ; *petals* deeply 2-cleft, twice as long as the sepals.—Among the most ornamental of our early summer flowers, scarcely less conspicuous with its delicate green leaves than its snow-white petals. The stems do not die down to the ground in the winter, as is the case

with most other herbaceous perennials ; but though dead
to all appearance, they send out delicate green tufts very

STELLARIA HOLOSTEA *and* S. MEDIA.

early in the year, so that the flowering stems, especially
in bushy places, seem to have made an unusually rapid
growth.—Fl. May, June.　Perennial.

　　3. *S. gramínea* (Lesser Stitchwort).— *Stem* nearly
erect, angular, smooth ; *leaves* very narrow, acute, smooth,
edged ; *flowers* in forked panicles ; *petals* very deeply
cleft, scarcely longer than the 3-nerved sepals.— Dry
heathy places.　Much smaller than the preceding in all

its parts, and distinguished at once by the very deeply divided petals, which are white, but not so showy.—Fl. June, July. Perennial.

4. *S. glauca* (Glaucous Marsh Stitchwort).—*Stem* nearly erect, angular, smooth ; *leaves* narrow, tapering, entire, glaucous ; *flowers* solitary on long axillary stalks ; *petals* very deeply 2-cleft, much longer than the 3-nerved sepals.—Marshy places. Resembling the preceding in habit, 6 — 12 inches high, but with larger flowers.—Fl. June—August. Perennial.

5. *S. uliginósa* (Bog Stitchwort).—*Stems* spreading, angular ; *leaves* broadly lanceolate, with a stiff tip, smooth ; *flowers* panicled ; *petals* deeply 2-cleft, shorter than the 3-nerved sepals, which are united at the base.—Boggy places, frequent. A slender plant, about a foot long, with very small white flowers.—Fl. June. Annual.

* The other British species of *Stellaria* are *S. némorum* (Wood Stitchwort), which grows in woods in the north, and is distinguished by its heart-shaped leaves : *S. cerastoides* (Alpine Stitchwort), a humble plant 4—6 inches high, with large white flowers, growing on the Scottish mountains ; and *S. scapígera* (Many-stalked Stitchwort), distinguished by its long flower-stalks, and growing only on the hills north of Dunkeld, and about Loch Nevis.

11. ARENARIA (*Sandwort*).

* *Leaves without stipules.*

1. *A. peploídes* (Seaside Sandwort).—*Leaves* sessile, egg-shaped, acute, fleshy, smooth ; *sepals* obtuse.— A low, succulent, marine plant, with creeping roots, forked stems, and leaves arranged in four rows. The flowers are small and white, and grow from the forks of the stem. The plant forms tangled masses on the sea-shore, and approaches in habit the Sea Milkwort (*Glaux maritima*).—Fl. July. Perennial.

ARENARIA PEPLOIDES (*Seaside Sandwort*).

2. *A. trinervis* (Three-nerved Sandwort).—*Leaves* egg-shaped, acute, the lower ones stalked, 3—5-nerved, fringed ; *flowers* solitary from the forks of the stem and axils ; *sepals* 3-nerved, the central nerve rough.—A weak, straggling, downy plant, about a foot long, approaching the Chickweed (*Stellaria média*) in habit, from which, however, it may be at once distinguished by its undivided petals.—Fl. May, June.   Annual.

3. *A. serpyllifolia* (Thyme-leaved Sandwort). — *Leaves* broadly egg-shaped, pointed, roughish, sessile ; *stem* repeatedly forked, downy ; *sepals* tapering, hairy. —A small shrub-like herb 2—6 inches high, with inconspicuous white flowers, common on dry banks and walls.   When growing near the sea, the stems are less

branched, and the leaves somewhat larger and more decidedly fringed.—Fl. June—August. Annual.

* To this group belong also *A. ciliata* (Fringed Sandwort), a small species growing on limestone mountains in Ireland : *A. Norvégica* (Norwegian Sandwort), discovered in the Shetland Islands, in 1837 ; *A. verna* (Vernal Sandwort), a small tufted plant with awl-shaped leaves, and comparatively large flowers, which grows in several parts of Scotland, and at the Lizard Point, Cornwall : *A. rubella* (Alpine Sandwort), a very rare species, confined to the summits of some of the Scotch mountains : *A. tenuifolia* (Fine-leaved Sandwort), a slender plant 4—6 inches high, with smooth forked stems and awl-shaped leaves, found in sandy fields in various parts of England and Scotland : and *A. fastigiata* (level-topped Sandwort), a rare Scotch species, with tufted flowers and leaves.

** *Leaves furnished with stipules.*

4. *A. rubra* (Purple Sandwort). — *Leaves* linear, somewhat fleshy, pointed with a minute bristle; *stipules* chaffy ; *stems* prostrate ; *seeds* rough.—A small branching annual, with purple flowers, common in sandy fields.— Fl. June—August. Annual.

5. *A. marína* (Seaside Sandwort).—*Leaves* semi-cylindrical without points ; *stipules* chaffy ; *stems* prostrate ; *seeds* smooth, flattened, bordered.—On the seashore ; common. Stouter and larger than the last, but nevertheless so nearly resembling it, that it is doubtful whether both may not belong to the same species, varied, however, by soil and situation. — Fl. June—August. Annual.

12. CERASTIUM (*Mouse-ear Chickweed*).

* *Petals not longer than the calyx.*

1. *C. viscósum* (Viscid Mouse-ear Chickweed.)—*Stems* hairy, viscid ; *leaves* oblong, tapering ; *flowers* in pani-

cles.—A common weed, with inconspicuous white flowers and straggling stems, which are generally covered with dust. The seed-vessels, when ripening, lengthen and become curved.—Fl. all the summer.　Perennial.

CERASTIUM VISCOSUM (*Viscid Mouse-ear Chickweed*).

* To this group belong *C. vulgátum* (Broad-leaved Mouse-ear Chickweed), which may be distinguished by its flowers being tufted : *C. semidecandrum* (Five-stamened Mouse-ear Chickweed), distinguished by the character to which it owes its name : *C. tetrandrum* (Four-stamened Mouse-ear Chickweed), distinguished in like manner : and several others to which botanists have given names, though undecided whether to call them species or varieties.

** *Petals longer than the calyx.*

2. *C. arvense* (Field Mouse-ear Chickweed).—*Leaves* narrow, tapering, downy ; *petals* twice as long as the sepals.—Gravelly and chalky fields; common.—Fl. June, July. Perennial.

3. *C. aquáticum* (Water Mouse-ear Chickweed).— *Lower leaves* stalked, *upper* sessile, heart-shaped, tapering to a point ; *capsule* opening with 5 2-cleft teeth.—The largest of the genus, and in habit approaching *Stellaria némorum* (Wood Stitchwort).—Wet places, but not general.—Fl. July, August. Perennial.

* To this group belong also *C. alpinum* (Alpine Mouse-ear Chickweed), a short plant with white silky leaves and large white flowers, frequent in the Highlands of Scotland : and *C. latifolium* (Broad-leaved Mouse-ear Chickweed), also a mountain plant, distinguished from the preceding by its leaves being covered with yellowish down, and by its solitary flowers. It is found on the mountains of Wales and Scotland.

CHERLERIA SEDOIDES (*Mossy Cyphel*).

13. CHERLERIA (*Cyphel*).

1. *C. sedoídes* (Mossy Cyphel).—A mountain plant with long roots and numerous densely tufted stems, which

scarcely rise above the ground, bearing crowded narrow leaves and solitary greenish flowers, which are generally without petals. — Highland mountains. — Fl. June — August. Perennial.

ORD. XV.—LINEÆ.—THE FLAX TRIBE.

Sepals 3—5 overlapping when in bud, persistent; *petals* equal in number to the sepals, twisted when in bud, falling off very soon after expansion ; *stamens* equal in number to the petals, and alternate with them, united at the base into a ring with small teeth between them ; *ovary* of about as many cells as there are sepals, and as many *styles* ; *capsule* approaching a globular form, tipped with the hardened base of the styles, each cell incompletely divided by a partition extending from the back inwards; *seeds* 1 in each imperfect cell, pendulous.—Herbaceous, rarely shrubby plants, with undivided leaves and remarkably fugacious petals, principally but not exclusively confined to Europe and the north of Africa. The flowers are in many cases highly ornamental, but the most striking feature of the Flax tribe is the toughness of the fibre contained in their stems, and the mucilaginous qualities of their seeds, which also yield considerable quantities of oil. One species, *Linum usitatissimum*, has for ages supplied the valuable article of clothing which takes its name, "linen," from the plant which produces it; lin-seed oil is obtained from the seeds of the same plant, and the meal of the seeds is valuable for poultices.

1. LINUM (Flax).—*Sepals* 5 ; *petals* 5 ; *capsule* 10-valved, and 10-celled. (Name from the Celtic *Lin*, thread.)

2. RADÍOLA (Flax-seed).—*Sepals* 4, connected below, 3-cleft ; *petals* 4 ; *capsule* 8-valved, and 8-celled. (" Named from *radius*, a ray, I presume in consequence of the ray-like segments of the calyx."—*Sir W. J. Hooker.*)

1. LINUM (*Flax*).

* *Leaves alternate.*

1. *L. perenne* (Perennial Flax).—*Leaves* very narrow, tapering to a sharp point ; *sepals* inversely egg-shaped, obtuse, obscurely 5-ribbed.—Chalky fields.—A slender

LINUM ANGUSTIFOLIUM *and* L. CATHARTICUM.

plant about a foot high, with wiry stems, very narrow sessile leaves, and elegant sky-blue petals, which are so fugacious as scarcely to bear being gathered.—Fl. June, July. Perennial.

2. *L. angustifolium* (Narrow-leaved Flax).—*Leaves* very narrow, tapering to a point ; *sepals* elliptical, pointed, 3-ribbed.—Sandy pastures in the southern and western counties ; common. Like the last, but distinctly

marked by its sharp-pointed sepals, and lighter blue flowers.—Fl. June, July. Perennial.

* To this group belongs *L. usitatissimum*, the flax of commerce, which, though not a native plant, is not unfrequently found in cultivated ground. It is distin-guished from the preceding by its somewhat broader and more distant leaves, by its stems being mostly solitary, instead of several from the same root, by its notched petals, and by its larger size.

** *Leaves opposite.*

3. *L. cathárticum* (Cathartic Flax).—*Leaves* oblong, broader at the base; *sepals* pointed.—Dry pastures; abundant.—Very different in size and habit from any of the preceding, rarely exceeding 6 inches in height, and bearing numerous small white flowers, which grow in panicles, and droop before expansion.—Fl. June, July. Annual.

RADIOLA MILLEGRANA (*Thyme-leaved Flax-seed*).

2. RADÍOLA (*Flax-seed*).

1. *R. millegrána* (Thyme-leaved Flax-seed).—The only species.—One of the most minute of British flowering

plants, never exceeding three inches in height, but repeatedly forked, and bearing a large number of small white flowers, which, as the plants generally grow many together, often prevent its being overlooked. — Damp heaths; not uncommon.—Fl. July, August. Annual.

ORD. XVI.—MALVACEÆ.—THE MALLOW TRIBE.

Sepals 5, more or less united at the base, valvate in bud, often enclosed in an outer calyx ; *petals* equalling the sepals in number, twisted when in bud ; *stamens* numerous, united by their filaments into a tube ; *ovary* formed of several carpels united in a radiate manner ; *styles* equal in number to the carpels, either distinct or united ; *capsules* (in all the British species) 1-seeded, arranged in a whorl round the styles. —A large and important family of herbaceous plants, shrubs, and trees, with divided alternate leaves, which are furnished with stipules and axillary flowers. They are most abundant in the tropical regions, where they form a large proportion of the vegetation, and gradually decrease in number towards the poles. According to Lindley, the number of species hitherto discovered amounts to about a thousand, all of which agree in containing a large quantity of mucilage, and being totally destitute of unwholesome qualities. In some species, this mucilage, extracted by boiling the plant, especially the root, is employed medicinally in allaying irritation, both external and internal. Some few are used as food. The bark of others affords an excellent substitute for hemp. The cotton of commerce is obtained from the appendage of the seeds of several species of Gossypium, a family belonging to this Order. As ornamental garden flowers, Malope, several species of Hibiscus, and the Hollyhock, are well known. The number of stove plants in cultivation is very great.

1. MALVA (Mallow).—*Styles* numerous ; *outer calyx* 3-leaved, *inner* 5-cleft. (Name from the Greek *malaché*,

soft, from the emollient properties of the mucilage which it contains.)

2. Lavatéra (Tree-Mallow).—*Styles* numerous ; *outer calyx* 3-lobed, *inner* 5-cleft. (Named in honour of the two *Lavaters*, friends of Tournefort.)

3. Althæa (Marsh-Mallow).—*Styles* numerous ; *outer calyx* 6—9 cleft. (Name from the Greek * áltho*, to cure, from its healing properties.

MALVA SYLVESTRIS (*Common Mallow*).

1. Malva (*Mallow*).

1. *M. sylvestris* (Common Mallow).—*Stem* ascending, or erect ; *root-leaves* kidney-shaped, with 7 acute lobes ; *fruit-stalks* erect ; *fruit* not downy, wrinkled.—Road-

sides and waste ground; common. A robust herbaceous plant, with large downy, lobed, but not deeply divided leaves, branched stems, and showy purple flowers. When the flowers first expand the plant is handsome, but as the season advances, the leaves lose their deep green hue, and the stems put on a ragged appearance. The whole plant is mucilaginous and emollient. The pollen is a beautiful object for the microscope, being (like that of many other plants in this Order) globular, and studded with minute glandular prickles.—Fl. June—August. Perennial.

2. *M. rotundifolia* (Dwarf Mallow).—*Stem* prostrate; *leaves* roundish, heart-shaped, with 5 shallow lobes; *fruit-stalks* bent down; *fruit* downy.—Waste places; not uncommon. Smaller than the last, and distinguished at once by its prostrate stems, and pale lilac flowers.—Fl. June—September. Annual.

3. *M. moscháta* (Musk-Mallow).—*Stem* erect; *root-leaves* kidney-shaped, deeply 5 or 7-lobed and cut; *stem-leaves* deeply 5-lobed, and variously cut into numerous narrow segments; *outer calyx leaves* very narrow; *fruit* hairy.—Hedges and borders of fields; not very common. Whole plant hairy, light green, with large handsome rose-coloured flowers, which are crowded towards the summit of the stem; the foliage emits a faint musky odour, especially in hot weather, or when drawn lightly through the hand. A white variety is not uncommon in gardens. Fl. July, August. Perennial.

* The above species agree in yielding a plentiful tasteless mucilage, which, in some rural districts, is highly valued for making poultices and cough mixtures. Children often amuse themselves with gathering and eating the unripe seed-vessels, which they call "cheeses:" they are insipid, but not unwholesome. The Common Mallow is frequently called, by country people, " Marsh Mallow," a name which properly belongs to *Althœa officinalis.*

LAVATERA ARBOREA (*Tree-Mallow*)

2. LAVATÉRA (*Tree-Mallow*).

1. *L. arbórea* (Sea Tree-Mallow).—A tall and hand-some plant, 6—12 feet high, with a thick, almost woody, stem, soft, downy, angular leaves, and abundance of purple flowers, resembling those of the common Mallow, but somewhat smaller, and of a deeper colour towards the

centre.—On sea-cliffs and insulated rocks on many parts of the coast.—Fl. July—October.   Biennial.

ALTHÆA OFFICINALIS (*Common Marsh-Mallow*).

3. ALTHÆA (*Marsh-Mallow*).

1. *A. officinalis* (Common Marsh-Mallow). — *Leaves* 3—5 lobed, soft and downy on both sides ; *flowers* several together.—Marshes, especially near the sea. Readily distinguished from any others of the Mallow tribe growing in Britain, by the numerous divisions of the outer calyx, by the hoary down which thickly clothes the stems and foliage, and by the numerous panicles of blush-coloured flowers.   The starry down is a beautiful object for the microscope.—Fl. August, September. Perennial.

I

* A second species of *Althæa*, *A. hirsúta* (Hispid. Marsh-Mallow), occurs in considerable abundance near, Cobham, Kent, but is not considered a native. It may be distinguished by its bristly stems and leaves, and by its flowers being solitary in the axils of the upper leaves.

Ord. XVII.—TILIACEÆ.—The Lime Tribe.

Sepals 4 or 5, valvate when in bud ; *petals* equalling the sepals in number, often with a little pit at the base, sometimes wanting ; *stamens* numerous ; *glands* 4—5, at the base of the petals ; *ovary* single, of 2—10 united, rarely distinct, *carpels ; style* 1, with as many *stigmas* as carpels ; *capsule* with one or more seeds in each cell. —The plants belonging to this natural order are mostly trees or shrubs. They have all a mucilaginous, wholesome juice, and many of them are remarkable for the toughness of the fibres of the inner bark. In a species of *Aristotelia* this is so strong, as to be converted into strings for musical instruments. One genus (*Córchorus*) supplies the Indians with fishing-lines and nets ; and the Lime or Linden-tree furnishes the material of which, in Russia, bast mats are made. Some genera produce edible berries, and the bony seeds of others are not uncommonly set in gold, and form handsome necklaces. In several instances the timber is employed for the most useful purposes. The name of the order is derived from *Tilia*, the Linden-tree, the only British genus.

1. Tilia (Lime).—*Sepals* 5, soon falling off ; *petals* 5, with or without a scale at the base outside ; *ovary* 5-celled ; *style* 1 ; *capsule* 1-celled not opening by valves, 2-seeded. (Name of uncertain origin.)

1. Tília (*The Lime or Linden-tree*).

1. *T. parvifolia* (Small-leaved Lime-tree).—*Leaves* obliquely heart-shaped, smooth on both sides, with the

exception of small tufts of downy hair beneath, scarcely longer than their stalks ; *peduncles* springing from a leafy *bract*, many flowered ; *capsule* brittle.—Peculiar interest attaches to the Linden-tree, from its having given a name to the immortal Linnæus. For a fuller account of this tree, the reader is referred to " The Forest Trees of Britain," vol. ii.—Fl. July. Tree.

TILIA PARVIFOLIA (*Small-leaved Lime-tree*).

* It is questionable whether any species of Lime is indigenous in Britain ; that just described has the best claim to be considered a native, but is not so well known as *T. Europœa*, the common Lime of avenues and parks. This species has flowers like those of the small-leaved Lime, but larger leaves, and a smooth capsule, the rind of which is tough. *T. grandifolia* (Large-leaved Lime-tree) has woody seed-vessels, which, as well as the leaves, are covered with short down.

ORD. XVIII.—HYPERICINEÆ.—THE ST. JOHN'S WORT TRIBE.

Sepals 4 or 5, not falling off, unequal, often fringed with black dots; *petals* of the same number as the sepals, unequal-sided, twisted when in bud, often bordered with black dots; *stamens* numerous (in Parnassia 5), united at the base into three or more sets; *ovary* single; *styles* 3—5; *fruit* a capsule or berry of several valves and cells, the *valves* curved inwards; *seeds* minute, numerous.— Herbs, shrubs, or trees, generally with opposite leaves marked with pellucid dots, and yellow flowers, inhabiting most parts of the world. Most of the species are aromatic and resinous, and some contain a yellow juice, which has been used in medicine. The only British genus strictly belonging to this order is that which gives it its name, *Hypericum.* The curious genus *Parnassia* is, however, by some botanists placed here, but, as is admitted, with questionable propriety.

1. HYPÉRICUM (St. John's Wort).—*Sepals* 5; *petals* 5; *stamens* numerous; *filaments* united into 3 or 5 sets; *styles* 3 or rarely 5; *capsule* 3-celled.—(Name, *hypericon,* the Greek name of the plant.)

2. PARNASSIA (Grass of Parnassus).—*Calyx* deeply 5-cleft; *petals* 5; *stamens* 5, with 5 fringed scales interposed; *stigmas* 4; *capsule* 1-celled, with 4 valves.— (Name from Mount Parnassus, but on what account is uncertain.)

1. HYPÉRICUM (*St. John's Wort*).

* *Styles* 5.

1. *H. calycinum* (Large-flowered St. John's Wort).— *Stems* single-flowered.—A low shrub, with oblong, blunt leaves and large yellow flowers, which are very handsome, especially before the anthers have shed their pollen. Common in gardens and shrubberies, and naturalized in several places.—Fl. July—September. Perennial.

** *Styles* 3, *sepals not fringed.*

2. *H. Androsœmum* (Common Tutsan).—*Stem* shrubby, two-edged ; *leaves* egg-shaped sessile ; *sepals* unequal ; *capsule* berry-like.—Woods and hedges, but not common, except in Devon and Cornwall. A handsome shrubby plant, 2—3 feet high, conspicuous with clusters of largish yellow flowers, and afterwards with glossy, berry-like capsules. The leaves have a strong resinous smell, which they retain for some time after drying.—Fl. July. Perennial. Shrub.

3. *H. quadrángulum* (Square-stalked St. John's Wort). —*Stem* herbaceous, erect, with 4 somewhat winged angles, branched ; *leaves* oblong, egg-shaped, with pellucid dots ; *sepals* erect, lanceolate.—Wet places, common, growing 1—2 feet high, with flat panicles of yellow flowers.— July, August. Perennial.

4. *H. perforátum* (Perforated St. John's Wort).— *Stem* herbaceous, erect, 2-edged ; *leaves* elliptic-oblong, copiously perforated with pellucid dots ; *sepals* erect, lanceolate, acute.—Woods and hedges, common ; easily distinguished from all the other British species of St. John's Wort by its decidedly 2-edged stem ; 2 feet high. —Fl. July, August. Perennial.

5. *H. dúbium* (Imperforate St. John's Wort).—*Stem* herbaceous, erect, 4-sided, with rounded angles ; *leaves* destitute of dots ; *sepals* reflexed, elliptical, blunt.— Mountainous places, not common : very like the last, but well distinguished by the above characters.—Fl. August. Perennial.

6. *H. humifúsum* (Trailing St. John's Wort).—*Stems* prostrate, somewhat 2-edged ; *leaves* oblong, obtuse, per- forated with pellucid dots ; *flowers* somewhat cymose ; *stamens* not more than 15.—Walls and gravelly banks, common ; 3—9 inches long, with several spreading slender stems and yellow flowers, which open most freely in the sunshine. Though usually placed in this group, the sepals are often fringed with black dots.— Fl. July. Perennial.

HYPERICUM PERFORATUM (*Perforated St. John's Wort*).

* * * *Styles* 3.   *Sepals fringed with glands.*

7. *H. montánum* (Mountain St. John's Wort).—*Stem* erect, round, smooth ; *leaves* oblong, sessile, smooth, with glandular dots near the margin ; *sepals* acute, fringed with shortly stalked glands.—Limestone hills, not uncommon ; growing about 2 feet high, and remarkable for the black fringe of its sepals, which at once distinguishes it from any of the preceding species.—Fl. July. Perennial.

8. *H. hirsútum* (Hairy St. John's Wort).—*Stem* erect, nearly round, downy ; *leaves* downy.—Woods, especially in a chalky or limestone soil ; about the same size as

the last, but well marked by its downy herbage.—Fl.
July. Perennial.

9. *H. pulchrum* (Small upright St. John's Wort).—
Stem erect, round, smooth ; *leaves* heart-shaped, embracing the stem, smooth ; *sepals* obtuse, fringed with sessile
glands ; *petals* fringed with glands.—Heaths and woods,
common. A slender plant, with scanty foliage and
golden-yellow flowers, which when in bud are stained
externally with red.—Fl. July, August. Perennial.

10. *H. elódes* (March St. John's Wort).—*Stem* creeping ; *branches* erect ; *leaves* roundish, and, like the stems,
densely clothed with shaggy down.—Spongy bogs, common. Flowers few, pale yellow, remaining open but a
short time. The plant may be detected at some distance by the hoariness of its foliage, and by the strong
and far from pleasant resinous odour which it emits,
especially in hot weather.—Fl. July, August. Perennial.

* The other British species of St. John's Wort are,
H. maculátum (Spotted St. John's Wort), which grows
in wet places, and may be distinguished from *H. quadrángulum* by its reflexed toothed *sepals*, and by purple
streaks and dots on the *petals* beneath : *H. linariifolium*
(Narrow-leaved St. John's Wort), which is nearly allied
to *H. humifúsum*, but has erect *stems*, and about 30
stamens ; found on the sea-coast of Devon and Cornwall,
very rare : and *H. barbatum* (Bearded St. John's Wort),
a doubtful native, but said to grow wild in Perthshire ;
it has *sepals* fringed with long-stalked *glands*, and *petals*
minutely fringed and dotted.

2. PARNASSIA (*Grass of Parnassus*).

1. *P. palustris* (Common Grass of Parnassus).—The
only British species.—Bogs, principally in the north.
An exceedingly elegant plant, 8—10 inches high, with
solitary cream-coloured flowers, beautifully veined. The
nectaries are fan-like scales, fringed with white hairs,
and terminating in yellow wax-like glands.—Fl. August
—October. Perennial.

PARNASSIA PALUSTRIS (*Common Grass of Parnassus*).

Ord. XIX.—ACERINEÆ.—The Maple Tribe.

Calyx divided into 5 parts (occasionally 4—9); *petals* of the same number; *stamens* about 8, inserted on a flattened ring beneath the ovary; *ovary* 2-lobed; *style* 1; *stigmas* 2; *fruit* 2-lobed, 2-celled, not bursting; *lobes* winged on the outside; *cells* 1—2 seeded.—Trees

ACER CAMPESTRE (*Common Maple*).

with opposite, stalked leaves, which are veined in a palmate manner. They are found only in the temperate regions of the northern hemisphere; several species abound in a sweet juice, which in North America is manufactured into sugar. — (See " Forest Trees of Britain," vol. i. p. 103.)

ACER PSEUDO-PLATANUS (*Greater Maple or Sycamore*).

1. ACER (Maple).—*Calyx* 5-cleft ; *petals* 5 ; *capsules* 2, each furnished with a long wing.—(Name from the Celtic *ac*, a point, on account of the hardness of the

wood, which was used for making spears and other sharp-pointed instruments.)

1. ACER (*Maple*).

1. *A. campestre* (Common Maple).—*Leaves* 5-lobed ; *lobes* bluntish, scarcely cut ; *clusters* of flowers erect.— Woods and hedges ; a small tree, with a very rugged corky bark, full of deep cracks.—Fl. May, June. Tree.

2. *A. pseudo-Plátanus* (Greater Maple or Sycamore). —*Leaves* 5-lobed ; *lobes* unequally serrated ; *clusters* of flowers drooping.—A large and handsome tree, introduced into England before the fourteenth century, and now completely naturalized. The name Sycamore was given to it by the older botanists, who erroneously believed it to be identical with the *Sycamore* or *Mulberry-fig* of Palestine, which it somewhat resembles in the size and form of its leaves.—Fl. May. Tree.

ORD. XX.—GERANIACEÆ.—GERANIUM TRIBE.

Sepals 5, not falling off, ribbed, overlapping when in bud ; *petals* 5 clawed, twisted when in bud ; *stamens* 10, sometimes alternately imperfect, generally united by their filaments ; *ovary* of 5 carpels placed round a long awl-shaped *beak ; styles* 5, united to the beak ; *stigmas* 5 ; *fruit* beaked, separating into 5 *capsules*, each of which is one-seeded, and terminates in the hardened style, which finally separates at the base and curls up, carrying the capsule with it.—An extensive tribe, consisting of herbaceous plants and shrubs with an aromatic or resinous flavour and astringent qualities, abounding at the Cape of Good Hope and in the temperate regions of the northern hemisphere. To the genus *Pelargonium* belong the innumerable varieties of handsome flowering plants which under the name of *Geraniums* are so ornamental as greenhouse or window flowers. The roots of one or two species are eatable, and some are used in medicine.

1. GERÁNIUM (Crane's-bill).—*Stamens* 10, 5 of which are alternately larger and have glands at the base ; *fruit* beaked, separating into 5 *capsules,* each with a long *awn* which is naked (not bearded on the inside).— (Name from the Greek *géranos,* a crane, to the beak of which bird the fruit bears a fancied resemblance.)

2. ERÓDIUM (Stork's-bill).—*Stamens* 10, of which 5 are imperfect ; *glands* 5, at the base of the perfect stamens ; *fruit* beaked, separating into 5 *capsules,* each with a long spiral *awn,* which is bearded on the inside. (Name from the Greek *eródion,* a heron, to the beak of which bird the fruit bears a fancied resemblance.)

1. GERANIUM (*Crane's-bill*).

* *Flower-stalks single flowered.*

1. *G. sanguíneum* (Bloody Crane's-bill).—*Root-leaves* nearly round, with 7 deeply cut lobes, each of which is 3-cleft, *stem-leaves* 5 or 3 lobed.—Limestone and magnesian rocks, not common. An exceedingly handsome plant with hairy stems, abundant foliage, and large, bright purple flowers. Fl. July—September. Perennial.

* * *Flower-stalks 2 flowered.*

2. *G. phœum* (Dusky Crane's-bill). — *Stem* erect ; *flowers* panicled ; *sepals* slightly pointed ; *capsules* keeled, hairy below, wrinkled above.—In woods and thickets, rare, but not an uncommon garden plant ; remarkable for the dingy, almost black hue of its flowers. Fl. May, June. Perennial.

3. *G. pratensé* (Meadow Crane's-bill).—*Stem* erect ; *leaves* palmate, 7-lobed ; *lobes* cut and serrated ; *stamens* smooth, tapering from a broad base ; *capsules* hairy all over ; *fruit-stalks* bent down.—The largest British species, growing in moist pastures ; about 2 feet high, with large and handsome purplish-blue flowers. Fl. June—August. Perennial.

4. *G. sylváticum* (Wood Crane's-bill). *Stem* erect ;
many flowered ; *leaves* palmate, 7-lobed ; *lobes* cut and
serrated ; *stamens* awl-shaped, fringed ; *capsules* keeled,
hairy ; *fruit-stalks* erect.—Woods and pastures, chiefly

GERANIUM PRATENSE (*Meadow Crane's-bill*).

in the north, rare. Distinguished from the last by its
smaller size, by its stamens being fringed half way up,
and by its capsules being most hairy about the keel.
—Fl. June, July. Perennial.

5. *G. Pyrenáicum* (Mountain Crane's-bill).—*Stem*
spreading, downy ; *root-leaves* kidney-shaped, 5—7

lobed ; *lobes* oblong, obtuse, 3-cleft and toothed at the end ; *petals* notched, twice as long as the pointed sepals. —Road-sides and meadows ; not common. 2—3 feet high. Well distinguished by the thick down on its stems and leaves, and by its numerous, rather small, purple flowers with cleft petals.—Fl. June, July. Perennial.

6. *G. Robertianum* (Herb Robert).—*Stem* spreading ; *leaves* ternate or quinate ; *leaflets* deeply cut, the segments with minute points ; *sepals* angular, hairy ; *capsules* wrinkled and hairy.—Road-sides and hedges ; very common. One of the most generally diffused and best known species, well distinguished by its red, hairy, succulent stems, and leaves which towards autumn acquire the same hue, and by its elegantly veined bright purple flowers. The scent of the whole plant is strong and unpleasant. Fl. all the summer. Annual.

7. *G. lúcidum* (Shining Crane's-bill).—Smooth and glossy. *Leaves* nearly round, 5-lobed ; *sepals* angular and wrinkled ; *capsules* with 3 keels, wrinkled.—Walls and stony places ; in many places. A beautiful little species, a few inches high, with small, rose-coloured flowers, and shining stems and leaves, which are generally tinged of a bright red.—Fl. all the summer. Annual.

8. *G. mollé* (Dove's-foot Crane's-bill).—Downy with soft hair. *Leaves* roundish, lobed and cut ; *petals* notched ; *capsules* wrinkled, not hairy ; *seeds* smooth.— Fields and waste places ; common. Easily distinguished from any of the preceding by its prostrate habit, downy herbage, and small light purple flowers.—Fl. all the summer. Annual.

9. *G. rotundifolium* (Round-leaved Crane's-bill).— Downy with soft hair. *Leaves* roundish, lobed, and cut ; *petals* entire ; *capsules* hairy, not wrinkled ; *seeds* dotted. —Fields and waste places ; not common, but perhaps often confounded with the last, which it much resembles in size and habit. — Fl. June—August. Annual.

10. *G. pusillum* (Small-flowered Crane's-bill). —
Downy with soft hair. *Leaves* roundish, lobed, and
cut; *petals* notched; *stamens* 5; *capsules* keeled, downy,
not wrinkled; *seeds* smooth.—Waste ground; common.
Resembling *G. molle* in habit, but smaller.—Fl. all the
summer. Annual.

11. *G. dissectum* (Jagged-leaved Crane's-bill).—*Stems*
spreading, hairy; *leaves* roundish, more or less hairy,
variously divided into numerous jagged, narrow segments; *sepals* with long points; *petals* notched; *capsules* scarcely wrinkled, hairy; *seeds* dotted.—Fields and
waste ground. Distinguished by its deeply cut, hairy,
not downy leaves, and the exceedingly short pedicels.
Flowers rose-coloured.—Fl. all the summer. Annual.

12. *G. columbínum* (Long-stalked Crane's-bill).—
Stems spreading, roughish with short hairs; *leaves*
deeply 5-lobed, the lobes cut into many long, narrow,
acute segments; *flower-stalks* very long; *sepals* with
long points; *capsules* smooth.—Waste ground; not very
common. Distinguished from the last by its larger
blueish rose-coloured flowers, which grow on very long
and slender stalks, and by its smooth capsules.—Fl.
June—August. Annual.

* Particular care should be taken when comparing
specimens with the above descriptions, to examine the
root-leaves; for the *stem*-leaves vary even on the same
plant to such a degree as to defy description.

2. ERÓDIUM (*Stork's-bill*).

1. *E. cicutarium* (Hemlock Stork's-bill). — *Stems*
prostrate, hairy; *stalks* many-flowered; leaves pinnate;
leaflets sessile, pinnatifid, cut.—Waste places, especially
near the sea; common. A straggling plant, with much
of the habit of the preceding genus, but distinguished
at first sight by its *pinnate* leaves, and *umbels* of lilac
(sometimes white) flowers, the petals of which soon fall
off.—Fl. all the summer. Annual.

2. *E. moschátum* (Musk Stork's-bill).—*Stems* pro-

strate, hairy; *stalks* many-flowered; *leaves* pinnate; *leaflets* nearly sessile, unequally cut; perfect *stamens*, toothed at the base.—Waste places, especially near the sea. Whole plant much stouter than the last, of a deeper

ERODIUM CICUTARIUM (*Hemlock Stork's-bill*).

green, somewhat clammy to the touch, and emitting when handled a strong scent of musk. Flowers bright purple.—Fl. all the summer. Annual.

3. *E. maritimum* (Sea Stork's-bill).—*Stems* prostrate,
hairy ; *stalks* 1—3 flowered ; *leaves* oblong, heart-shaped,
variously lobed and notched ; *petals* minute or wanting.
—Waste places near the sea; not uncommon in the
West of England. Whole plant roughish with minute
hairs, and sending out several leafy stems, which lie
remarkably close to the ground ; the leaves are not pinnate, as in the other British species, and the flowers are
rarely found with petals. Like many other sea-side
plants, it is not unfrequently met with in inland mountainous districts, occurring plentifully on Dartmoor, in
Devonshire, many miles from the sea.—Fl. all the
summer. Perennial.

* The beaks attached to the capsules of the Stork's-
bills become spirally twisted when ripe, often springing
to a considerable distance from the parent plant. They
are furnished on the inner side with long elastic bristles ;
and being hygrometric, uncurl when moistened. The
combined action of the beak and bristles thus gives to
the seed the power of locomotion at every change in the
moisture of the atmosphere. A twisted capsule, if
moistened and laid on a sheet of paper, will, in its effort
to straighten itself, soon crawl an inch or more away
from the spot on which it was laid.

Ord. XXI.—BALSAMINEÆ.—Balsam Tribe.

The flowers of this order are so exceedingly irregular,
that it is almost impossible to define the characters
without employing terms which would be out of place
in a work which professes to give merely a popular
description of British wild flowers. The following
description, however, of the only British species belonging
to the Balsam tribe, will serve to identify any others
which are likely to fall in the reader's way.—An annual
succulent plant, much swollen at the joints, with a solitary branched stem, and egg-shaped, deeply serrated
leaves. From the axil of each of the upper leaves

K

proceeds a flower-stalk, taking a horizontal direction, and hiding itself beneath the leaf. Each flower-stalk bears about four drooping flowers, which expand one at

IMPATIENS (*Balsam*).

a time, and last a very little while. The calyx consists of two coloured, nearly round concave sepals, with an oblique point ; within these, on the side of the flower nearest to the stem, is inserted a horn-like petal or sepal, for botanists are undecided which to call it, wide at the mouth and curved downwards at the extremity ; on each side of this is a large wavy petal unequally lobed, the largest lobe next the spur, the smaller being easily separable, and having the appearance of a distinct petal. Opposite the stem is a very broad, wavy petal ; and at its base are 5 stamens with short filaments, united just beneath the anthers into a ring, and enclosing a 5-celled ovary. The sepals and petals soon fall off, when the ovary enlarges to a 5-celled, 5-valved capsule, externally resembling a cylindrical, strongly ribbed pod. As the seeds approach towards maturity, the valves of the capsule acquire an extraordinary elastic power, and, if touched, instantaneously curl into a spiral form, and spring with considerable force many feet from the plant, dropping the seeds in the way.

Only one of the Balsam tribe is a native of Europe, but in India they are very numerous, and several are cultivated in British gardens.

1. IMPATIENS (Balsam).—Characters described above. The name, signifying *impatient,* was given from the sudden curling of the valves of the capsule when touched. The only English species, *I. Noli-me-tangere (do not touch me),* is an elegant plant, 1—2 feet high, with large flowers of a delicate yellow, beautifully spotted with orange colour.—It grows in moist shady woods in York-shire and Westmoreland.—Fl. July, August. Annual.

ORD. XXII.—OXALIDEÆ.—THE WOOD-SORREL TRIBE.

Sepals 5, persistent ; *petals* 5, equal, often united at the base, twisted when in bud ; *stamens* 10, with the *filaments* generally combined at the base, the 5 outer shortest ; *ovary* 3—5 celled ; *styles* 3—5 ; *capsule* 3—5 celled, with as many, or twice as many, valves ; *seeds* enclosed each in an elastic case, which curls back when the fruit is ripe, and throws the seed to a distance.—

Herbaceous plants, shrubs, or trees, remarkable for the quantity of oxalic acid contained in their foliage, on which account some species are used as sauces, salads, or even pickles. *Oxális crenáta* was a few years back recommended to be extensively cultivated for the sake of its tuberous roots, which, it was said, would form an excellent substitute for the potato. It has, however, proved to be of little value. Some species have sensitive leaves, and a few are used in medicine.

1. OXÁLIS (Wood-Sorrel).—*Sepals* 5, united below; *petals* 5, often united below; *stamens* united by the base of their filaments; *styles* 5; *capsules* 5-celled, angular. (Name from the Greek *oxys*, sharp or acid, from the acidity of the leaves.)

OXALIS ACETOSELLA (*Common Wood-Sorrel*).

1. OXÁLIS (*Wood-Sorrel*).

1. *O. acetosella* (Common Wood-Sorrel). — *Leaves* all springing directly from the root, ternate, hairy; *scape*

with two bracts about the middle, single flowered ; *root*
toothed.—An elegant little plant, with delicate drooping
clover-like leaves, and white or lilac veined *flowers*,
growing abundantly in woods and shady places. The
leaves, though not so sensitive as some foreign species,
fold together at night. This plant is supposed by many
to be the true shamrock which was used by St. Patrick
to illustrate familiarly the doctrine of the Trinity.—Fl.
May, June. Perennial.

*O. corniculata (Yellow Wood-Sorrel) has a prostrate
stem and small yellow flowers. It occurs only in the
West of England, and is by no means so pretty a plant
as the last.

SUB-CLASS II.

CALYCIFLORÆ.

Sepals distinct or united ; *petals* distinct; *stamens*
inserted on the *calyx*, or close to its base.

ORD. XXIII.—CELASTRINEÆ.—SPINDLE-TREE TRIBE.

Sepals 4—5, imbricated when in bud, inserted on a
fleshy disk ; *petals* equal in number to the sepals ;
stamens equal in number to the petals, and alternate
with them ; *ovary* sunk in the disk, 2—5 celled ; *fruit*
either a capsule of 2—5 cells, opening with valves, or
berry-like ; *seeds* often wrapped in a covering distinct
from the capsule (called an *arillus*).—A rather large
number of plants are included in this order, but not
many of great interest. They are natives of the warmer
parts of Europe, North America, and Asia, and a great
number inhabit the Cape of Good Hope. A few also
occur in Chili, Peru, and New Holland. Many of them

possess an acrid, stimulant principle. The green leaves of one species are said to be eaten by the Arabs to produce watchfulness, and a sprig of it is believed to be

EUONYMUS EUROPÆUS (*Common Spindle-Tree*).

a protection from the plague to the person who carries it. The only British species, the Spindle-tree, is most remarkable for its pink lobed seed-vessels, which, in autumn, render the tree a conspicuous object. One species of *Celastrus* (the genus from which the order takes its name) is said to inflict very painful wounds. The English name, *Spindle-tree*, is derived from the use made of its very compact wood.

1. EUÓNYMUS (Spindle-tree).—*Capsule* 3—5 angled, with 3—5 *cells* and *valves; seeds* solitary in each cell, coated with a fleshy *arillus*. (Name from *Euónymé*, the mother of the Furies, in allusion to the injurious properties of the fruit.)

1. EUÓNYMUS (Spindle-Tree).

1. *E. Europœus* (Common Spindle-Tree). — *Petals* usually 4, oblong, acute ; *stamens* usually 4 ; *branches* angular, smooth ; *leaves* broadly lanceolate, minutely serrated.—A hedge and wood shrub, well marked by its clusters of small greenish *flowers*, glossy *leaves*, green *bark*, and, above all, by its deeply lobed *seed-vessels*, which, when ripe, are rose-coloured, and on opening disclose the *seeds* curiously wrapped in a scarlet *arillus*. The wood, like that of the wild Cornel and Guelder-rose, is much used for making skewers. (*See* " Forest Trees of Britain," vol. ii.)—Fl. May. Shrub.

ORDER XXIV.—RHAMNEÆ. — BUCKTHORN TRIBE.

Calyx 4—5 cleft, valvate when in bud ; *petals* 4—5, inserted into the throat of the calyx; *stamens* 4—5, opposite the petals ; *ovary* superior, or half superior, 2—4 celled, surrounded by a fleshy *disk; fruit* either fleshy and not bursting, or dry, and separating in three divisions ; *seeds* several.—Trees or shrubs, with simple *leaves*, minute *stipules*, and small greenish *flowers*. Some species of *Zizyphus* produce the jujube, well known in this country as a sweetmeat. *Z. Lotus* is

famous for being the plant which afforded food to the
ancient Lotóphagi, or Lotus eaters. Homer states that it
was so delicious, that whatever stranger once tasted it,

RHAMNUS CATHARTICUS (*Common Buckthorn*).

immediately forgot his friends and native country, and
desired only to dwell within reach of it. It is a prickly
shrub, and bears an abundance of purplish berries, of

the size of sloes, and containing large stones. The pulp is mealy, and of a delicious flavour. Under the name of *seedra*, or *sadr*, it still affords food to the Arabs, who separate the pulp from the stone by gently pounding the fruit in a mortar, and either convert it into a kind of bread at once, or lay it by for winter use. A kind of wine is also made from the fruit, but this will not keep for more than a few days. Mungo Park, Dr. Shaw, and

RHAMNUS FRANGULA (*Alder Buckthorn*).

other travellers, found the tree in abundance in many of the sandy parts of Arabia; and the latter states that the fruit, called *nabk*, is regularly exposed for sale in the markets of Barbary. *Z. spina-Christi*, and *Paliurus aculeatus*, prickly shrubs, common in the East, are severally believed by many persons to have furnished our Blessed Saviour's crown of thorns. Only two plants of this tribe are indigenous to Britain, and belong to the genus *Rhamnus* ; their berries are medicinal, but too violent in their effects to be used with safety.

1. RHAMNUS (Buckthorn).—*Calyx* vase-like, 4—5 cleft ; *petals* 4—5 (sometimes wanting) ; *stamens* 4—5, inserted with the petals into the throat of the calyx ; *berry* 2—4 celled. (Name from the Greek *rhamnos*, a branch.)

1. RHAMNUS (*Buckthorn*).

R. cathárticus (Common Buckthorn).—*Branches* terminating in thorns ; *flowers* 4-cleft, diœcious (stamens and pistils on separate plants) ; *leaves* egg-shaped, sharply serrated ; *berry* 4-seeded.—A spreading shrub with dense branches of small green *flowers*, which are succeeded by black, powerfully cathartic *berries*. These, if gathered before they are ripe, yield a yellow dye ; when ripe, they form, if mixed with gum-arabic and lime-water, the green colour known under the name of Bladder-green.—Woods and thickets.—Fl. May. Shrub.

R. Frángula (Alder Buckthorn).—*Branches* without thorns ; *flowers* 5-cleft, all perfect ; *leaves* entire, smooth ; *berry* 2-seeded.—A rather slender shrub, 6—10 feet high, with smooth blackish *branches*, deep green *leaves*, and small greenish *flowers*, which are not so densely tufted as in the last.—Woods and thickets.—Fl. May. Shrub.

ORDER XXV.—LEGUMINOSÆ.—PEA AND BEAN TRIBE.

Calyx 5-cleft, with the odd lobe in front ; *petals* 5, papilionaceous ; *stamens* 10, their filaments either united into a tube, or forming 2 sets of 9 and 1 ; *ovary, style,* and *stigma*, single ; *seed-vessel* a 2-valved, sometimes imperfectly jointed *pod*, or *legume ; seeds* on the upper seam of the pod-valves.—A highly interesting order of plants, containing as many as 6,500 species, which vary in size from minute herbs to vast trees, with trunks upwards of 80 feet in circumference. In structure, properties, colour of flowers, and range of growth, they vary scarcely less than in dimensions ; they are found in all parts of the known world, except St. Helena, and another remote island ; many species, under the general name of pulse, afford the most nutritious food—for example, Peas, Beans, and Lentils ; others supply valuable fodder for cattle, as Clover, Vetches, and Lucerne ; Rosewood, Logwood, and Acacia offer examples of timber ; Gum Arabic, Catechu, Senna, Kino, Liquorice, Balsam of Tolu, and Tamarinds, are the products of other species ; Tonka Bean and Balsam of Peru are well known perfumes ; several species of *Indigófera* afford the valuable article of commerce, Indigo ; and in Persia and Bokhara, a tree called Camel's Thorn produces abundance of *Manna*, which in those countries is an important article of food. Other species possess medicinal properties of various kinds ; not a few are poisonous ; and it is worthy of remark that some, the seeds of which are eminently nutritious, have properties of an opposite nature residing in other parts of the plant. The roots of the Kidney Bean, for instance, are dangerously narcotic. The dry pulp which encloses the seeds of the Carob-tree is by some supposed to have been the food of John the Baptist in the wilderness, on which account it is called Locust-tree, and St. John's Bread. The Locust-trees of America (different from the eastern tree) attain an immense size. Martius represents a scene in Brazil, where some trees of this kind occurred of such enormous dimensions, that 15 Indians with outstretched arms could only just embrace one of them. At the bottom they were 84 feet in circumference, and 60 feet where the boles became cylindrical. According to one esti-

mate, they were 2,052 years old, while another carried
it up to 4,104. Many plants belonging to the *Mimosa*
group display peculiar irritability in their pinnate
leaves. This is particularly the case with *M. sensitiva*
and *M. pudica*, which are commonly called sensitive
plants. Almost all the plants of the order which have
compound leaves fold them together during the night.
In some foreign species of Leguminosæ the legume loses
its characteristic form, and assumes the appearance of
a drupe, the papilionaceous form of the flower remaining ; in others the petals lose the papilionaceous arrangement ; but the seed-vessel retains the form of a legume.
All the British species, however, are decidedly papilionaceous, and the principal varieties of form in the pod
are those of the Bird's-foot and others, where it is imperfectly jointed, and in Medick, where it is often
spirally twisted, so as to resemble a snail-shell. The
number of British species amounts to nearly seventy, of
which two species of Furze, three of *Genista*, and one of
Broom, are shrubs ; the rest are herbaceous. For convenience of reference, these are divided into groups.

Group I.—LOTEÆ.—*The Lotus Group.*

*Legume not jointed ; leaves simple, of 3 leaflets, or pin-
nate with an odd one.*

* *Leaves simple, or of 3 leaflets; stamens all united by
their filaments.*

1. ULEX (Furze).—*Calyx* of 2 sepals, with 2 minute
bracts at the base ; *legume* swollen, few-seeded, scarcely
longer than the calyx. (Name from the Celtic *ec* or *ac*,
a prickle.)
2. GENISTA (Green-weed).—*Calyx* 2-lipped, the upper
lip 2-cleft, the lower with 3 teeth ; *standard* oblong;
style awl-shaped ; *legume* swollen, or flat. (Name from
the Celtic *gen*, a shrub ; *Planta Genista* originated the
distinctive name of the Plantagenet family.)

3. Cýtisus (Broom).—*Calyx* 2-lipped, the upper lip with 2, the lower with 3 teeth ; *standard* broadly ovate; *style* thickened upwards ; *legume* flat, many-seeded. (Name, the Greek name of the plant.)

4. Onónis (Rest-harrow).—*Calyx* 5-cleft, its segments very narrow ; *keel* beaked ; *style* thread-like ; *legume* swollen, few-seeded. (Name from the Greek *onos*, an ass, by which animal the plant is eaten.)

** *Leaves of 3 leaflets ; stamens in 2 sets of 9 and 1.*

5. Medicágo (Medick).—*Legume* sickle-shaped, or spirally twisted. (Name of Greek origin, and denoting that some plant of the family was introduced from Media.)

6. Melilótus (Melilot).—*Calyx* with 5 nearly equal teeth ; *petals* distinct, soon falling off ; *legume* of few seeds, longer than the calyx. (Name from *mel*, honey, and *lotus*, the plant so called.)

7. Trifólium (Trefoil).—*Calyx* with 5 unequal teeth; *petals* combined by their claws, and persistent ; *legume* of few seeds, concealed in the calyx. (Name from *tria*, three, and *folium*, a leaf, each leaf being composed of 3 leaflets.) N.B.—In *T. ornithopodioïdes* the legume is 8-seeded, and twice as long as the calyx.

8. Lotus (Bird's-foot Trefoil).—*Calyx* with 5 nearly equal teeth ; *legume* cylindrical, many-seeded, and imperfectly many-celled. (Name from the Greek *lotos*.)

*** *Leaves pinnate, with a terminal leaflet.*

9. Anthyllis (Lady's Fingers).—*Stamens* all united by their filaments ; *calyx* inflated, 5-toothed ; *legume* enclosed in the calyx. (Name from the Greek *anthos*, a flower, and *ioulos*, down, from the downy calyx.)

10. Oxýtropis.—*Stamens* in 2 sets, 9 and 1 ; *keel* of the *corolla* pointed ; *legume* more or less perfectly 2-celled. (Name from the Greek *oxys*, sharp, and *tropis*, a keel.)

11. Astrágalus (Milk Vetch).—*Stamens* in 2 sets,

9 and 1 ; *keel* of the *corolla* blunt ; *legume* more or less
perfectly 2-celled. (Name from the Greek *astrágalos,*
a pastern bone, from the knotted form of the root of the
plant to which the name was originally given.)

Group II.—VICIEÆ.—*The Vetch Group.*

Legume not jointed; stamens in 2 sets, 9 and 1 ; leaves
pinnate, terminating in a tendril or short point.

12. VÍCIA (Vetch).—*Calyx* 5-cleft ; *style* thread-like,
with a small tuft of down beneath the stigma ; *leaves*
with tendrils. (Name originally derived, according to
Théis, from *gwig,* Celtic ; *wicken,* in German ; *bikion,* in
Greek ; *vesce,* in French.—*Hooker.*)

13. ERVUM (Tare).—*Calyx* 5-cleft ; *style* thread-
like ; *stigma* downy all over ; *leaves* with tendrils.
(Name from the Celtic *erw,* a ploughed field, its common
place of growth.)

14. LÁTHYRUS (Vetchling). — *Calyx* 5-cleft ; *style*
flattened on the upper side, downy beneath the stigma ;
leaves with tendrils, except in *L. Nissolia.* (Name from
the Greek *lathyros,* a plant so called.)

15. ÓROBUS (Bitter Vetch).—*Calyx* 5-cleft, swollen at
the base, oblique at the mouth, its upper *segments* deeper
and shorter ; *style* flattened on the upper side, downy
beneath the stigma ; *leaves* ending in a short point.
(Name from the Greek *oro,* to stimulate, and *bous,* an
ox, from its nutritive properties.)

Group III.—HEDYSAREÆ.—*The Joint-Vetch Group.*

Legume divided into 1-seeded joints or cells ; leaves pin-
nate, with an odd leaflet.

* *Flowers simple, in umbels.*

16. ORNÍTHOPUS (Bird's-foot). — *Legume* curved,
divided into many equal-sided joints, each of which

contains a seed ; *keel* small, obtuse. (Name from the Greek *ornis*, a bird, and *pous*, a foot, to which the tufts of seed-vessels bear a singular resemblance.)

17. HIPPOCRÉPIS (Horse-shoe Vetch).—*Legume* composed of numerous crescent-shaped joints, so that each legume has many deep notches on one side ; *keel* narrowed into a beak. (Name from the Greek *hippos*, a horse, and *crepis*, a shoe, from the form of the joints of the seed-vessel.)

** *Flowers in racemes.*

18. ONÓBRYCHIS (Saint-foin).—*Legume* straight, 1-celled, 1-seeded, not opening, the lower edge fringed or winged. (Name from the Greek *onos*, an ass, and *brycho*, to bray, it being supposed that the smell excites braying.)

1. ULEX (*Furze*).

1. *U. Europæus* (Common Furze, Gorse, or Whin).—*Bracts* ovate, not adhering closely to the calyx ; *branches* copiously beset with branched thorns.—A much branched, spreading shrub, almost leafless, except in its seedling state, when the leaves are composed of 3 narrow, soft leaflets. It attains maturity in about four years, but in sheltered places continues to grow until it reaches a height of from 12 to 18 feet. Its natural habit is, however, to grow on dry, exposed commons, which, in its flowering season, it covers with a gorgeous sheet of golden blossoms, entirely concealing its somewhat unsightly branches. Perhaps no plant is so broadly characteristic of English scenery, and the English climate, as " Yellow Whin." It does not thrive in hot countries ; and if removed to a much colder climate, pines and dies ; it is rare even in the Highlands of Scotland. The seed-vessels burst elastically in hot weather with a crackling noise, scattering the seeds on all sides. The calyx-teeth of this species are so closely

united as to be scarcely visible.—Fl. February—June. Shrub.

* A variety has been found in Ireland, which does not flower freely, and also differs from the common form of the plant in having a soft and succulent, instead of a rigid, habit. This variety has been cultivated with success as fodder for sheep and oxen.

ULEX EUROPÆUS (*Common Furze*).

2. *U. nanus* (Dwarf Furze).—*Calyx-teeth* spreading ; *bracts* minute, closely pressed to the calyx.—A very distinct species from the last, with which, however, it is sometimes confounded. It may readily be distinguished

by the above characters, by being smaller in all its parts, and by the spreading wings of its flowers, which, moreover, usually appear at the same season with the

GENISTA TINCTORIA (*Dyer's Green-weed, Woad-waxen*).

Heath, a plant with which it loves to intertwine its rigid branches.—Fl. August—November.   Shrub.

L

2. Genista (*Green-weed*).

1. *G. Ánglica* (Needle Green-weed, or Petty Whin).
—*Stems* thorny and leafless below ; *leaves* narrow,
smooth ; *legumes* smooth, inflated.—Not uncommon on
heaths and moors. A low shrub, about a foot high,
with reclining tough stems, which are armed at intervals with groups of slender, very sharp thorns. The
upper branches are destitute of thorns, and produce leafy
clusters of yellow flowers, which (like some other yellow
flowers belonging to this Natural Order) are remarkable
for turning green in drying.—Fl. May, June.   Shrub.

2. *G. tinctoria* (Dyer's Green-weed, Woad-waxen).
— Thornless ; *leaves* narrow, acute, nearly smooth ;
flowers in clusters ; *legumes* flattened, smooth.—A low
shrub, about a foot high, with tough stems, bright-green
foliage, and yellow flowers on short stalks.   It grows
in heathy places and fields, and is used to dye yarn of a
yellow colour.—Fl. July, August.   Shrub.

3. *G. pilósa* (Hairy Green-weed).—Thornless ; *leaves*
narrow, obtuse, the lower ones often inversely heart-
shaped, silky beneath ; *flowers* axillary, on short stalks ;
legumes downy.—Heathy places, rare.   A low shrub,
with prostrate stems, which are gnarled and much
branched, and small yellow flowers.—Fl. May, and again
in autumn.   Shrub.

3. Cýtisus (*Broom*).

1. *C. scoparius* (Common Broom).—The only British
species, well distinguished by its numerous, slender, erect
branches, with small scattered leaves of 3 leaflets, and
large yellow flowers.   The legumes when ripe are nearly
black, and hairy at the margin.—Fl. June.   Shrub.

CYTISUS SCOPARIUS (*Common Broom*).

ONÓNIS ARVENSIS (*Common Rest-harrow*).

4. ONÓNIS (*Rest-harrow*).

1. *O. arvensis* (Common Rest-harrow).—*Stem* shrubby, hairy; *leaflets* oblong; *flowers* axillary; *calyx* much shorter than the corolla.—Barren, sandy places, common, especially near the sea. A very variable plant, sometimes spreading on the ground, and rooting at the joints; at other times forming a small leafy bush. The roots are tough and very long, hence the English name. The branches often terminate in thorns; the leaves are viscid; the flowers of a bright rose-colour, and handsome. According to some botanists, several species are included under one name.—Fl. all the summer. Perennial.

* *O. reclináta* (Small Spreading Rest-harrow), is a

herbaceous species which, as a British plant, occurs only near Tarbert, Galloway.

5. MEDICÁGO (*Medick*).

1. *M. lupulína* (Black Medick, or Nonsuch).—*Leaflets* inversely egg-shaped, finely toothed ; *stipules* scarcely notched ; *flowers* in dense oblong heads ; *legumes* rugged, 1-seeded, kidney-shaped.—In pastures, common. A herbaceous plant, with the habit of some of the smaller Clovers, from which it may be distinguished by its legumes not being enclosed within the calyx. *Flowers* yellow ; *legumes* black, not spirally curved.—Fl. June— August. Annual.

MEDICAGO MACULATA (*Spotted Medick*).

2. *M. maculáta* (Spotted Medick).—*Leaflets* inversely heart-shaped ; *stipules* toothed ; *flowers* 2—4 together ; *legumes* spirally twisted into a prickly ball ; *prickles* curved.—In the south and west of England, not uncommon. The leaflets are remarkable for having in their

centre a purple, heart-shaped spot. In Cornwall this plant, under the name of *Spotted Clover*, is considered very injurious to pasturage.—Fl. yellow, June—September. Annual.

3. *A. mínima* (Little Bur-Medick).—*Leaflets* inversely egg-shaped, downy ; *stipules* scarcely toothed ; *flowers* 2—4 together ; *legumes* spirally twisted into a prickly ball ; *prickles* hooked.—Sandy places, rare.—Fl. yellow, June, July. Annual.

MELILÓTUS OFFICINALIS (*Common Yellow Melilot*).

4. *A. denticulata* (Toothed Medick). —*Leaflets* inversely heart-shaped, smooth ; *stipules* jagged ; *flowers*

2—4 together ; *legumes* loosely spiral; *prickles* hooked.
—Very rare.—Fl. yellow, April—June. Annual.

* Besides the above, two other species of *Medicágo*
are occasionally found, which are supposed to have
escaped from cultivation. *M. satíva* (Lucerne), which
has purple flowers, and *M. falcáta* (Sickle Medick), an
allied species with yellow flowers. In these two the
leaflets are oblong and toothed, and the legumes are
slightly twisted.

6. MELILÓTUS (*Melilot*).

1. *M. officinális* (Common Yellow Melilot).—*Stem*
erect ; *leaflets* narrow, egg-shaped, serrated ; *flowers* in
1-sided clusters ; *petals* equal in length ; *legumes* 2-
seeded, wrinkled.—A branched herbaceous plant, 2 to 3
feet high, with light-green foliage, and small yellow
flowers, not uncommon in waste places. This plant,
while drying, smells like Woodruff, or new hay.—Fl.
June—August. Annual.

* *M. leucantha* (White Melilot), a much less common
plant, differs from the preceding in having white flowers,
in which the standard is longer than the wings and keel.

7. TRIFOLIUM (*Trefoil*).

* Legumes with several seeds.

1. *T. repens* (White or Dutch Clover).—*Flowers* in
roundish heads, stalked, finally bent back ; *legumes*
4-seeded ; *stems* creeping. — Abundant in meadows,
where it forms excellent pasture. The flowers are white,
sometimes tinged with pink, and fragrant; on the sandy
sea coast they appear early in spring, in which situations the foliage is stunted and scanty. The leaflets
often have a white line near the middle, and sometimes
a dark spot as well. In a variety commonly cultivated
in gardens, under the name of Shamrock, nearly the

whole of the centre of each leaflet is tinged with dark
purple.   The real Shamrock is this species, and, perhaps, any other " 3-leaved grass" which grows in similar situations.   Much discussion about the identity of

TRIPOLIUM STRICTUM, SUBTERRANEUM, REPENS, *and* ARVENSE.

Shamrock might have been saved, by recollecting that
St. Patrick's day falls at a season when the botanical
characters of the trefoils are scarcely developed, and that
the devotees of that Saint can scarcely be expected to
possess much botanical knowledge.   Some antiquarians

contend that, as Ireland was a well-wooded country in
St. Patrick's time, the Saint very probably selected a
leaf of Wood-Sorrel (*Oxális acetosella*) to illustrate the
Doctrine of the Trinity.—Fl. through the summer.
Perennial.

2. *T. ornithopodioídes* (Bird's-foot Trefoil).—*Flowers*
2—3 together ; *legumes* 8-seeded, twice as long as the
calyx.—Dry, sandy places, not common. A small plant,
with spreading prostrate branches and small reddish
flowers, erroneously retained in the present genus.

** *Legumes* 1 *or* 2-*seeded ; standard falling off, or
    remaining unaltered ; calyx not inflated.*

3. *T. pratensé* (Purple Clover).—*Flowers* in dense,
roundish oblong heads ; *calyx* hairy, its bristle-like
divisions half as long as the corolla ; *stipules* broad,
terminating abruptly in a bristle-point ; *leaflets* broad,
notched, or entire.—The common Clover of meadows,
where it forms a valuable part of the hay crop. The
long tubes of the corolla abound in honey, on which
account they are often called by children Honeysuckles.
—Fl. all the summer. Perennial.

4. *T. médium* (Zigzag Clover).—*Flowers* in loose,
stalked, round heads ; *calyx-teeth* bristle-like, the 2
upper shortest ; *stipules* narrow, tapering to a point ;
leaflets narrow, elliptical ; *stem* zigzag.—Dry pastures
and bushy places ; common. Not unlike the last, but
well distinguished by its slenderer and more erect habit,
and especially by its narrow *leaflets*, and tapering, not
abrupt, *stipules*. It thrives better than the preceding
in dry soils, but is said to contain not more than half
as much nutritious matter.—Fl. July, August. Perennial.

5. *T. ochroleucum* (Sulphur-coloured Trefoil). —
Flowers in dense, stalked, terminal heads, which are at
first hemispherical, afterwards egg-shaped ; *calyx-teeth*
awl-shaped, the lower one much the longest ; *lower
leaflets* heart-shaped, upper oblong.—Not unfrequent in

the eastern counties, growing a foot or more in height. The whole plant is downy; the flowers are cream-coloured, turning brown as they fade. The lower leaves are on very long stalks.—Fl. July, August. Perennial.

6. *T. arvensé* (Hare's-foot Trefoil).—*Flowers* in terminal oblong heads, which are soft with downy hair; *calyx-teeth* hairy, much longer than the corolla; *stem* branched, erect.—A very distinct species, common in sandy places, especially near the sea. The peculiarly soft heads, which are nearly cylindrical, and in which the pale-pink flowers are almost concealed, at once distinguish this from any other British species. — Fl. July—September. Annual.

7. *T. marítimum* (Teasel-headed Trefoil).—*Flowers* in terminal roundish heads; *calyx-teeth* broad, pointed, and rigid, shorter than the corolla, finally becoming enlarged and spreading; *stipules* awl-shaped, very long. —Salt marshes; not common. Stem spreading; flowers pink, small.—Fl. June—July. Annual.

8. *T. scabrum* (Rigid Trefoil).—*Flowers* in dense prickly heads, which are both terminal and axillary; *calyx-teeth* unequal, very rigid, finally spreading; *stems* prostrate.—Barren places, especially near the sea. A small plant, with inconspicuous whitish flowers, and remarkable only for its prickly calyces, especially when in fruit.—Fl. June, July. Annual.

9. *T. striatum* (Soft Knotted Trefoil).—*Flowers* in downy heads, which are both terminal and axillary; *calyx* rigid, furrowed, with straight, unequal, awl-shaped teeth, swollen when in fruit; *stems* ascending.—Barren places, especially near the sea; common. Larger than the last, with which it is often found growing. The whole plant is covered with more or less silky hairs; the flowers are light purple, and the calyx is striated with prominent nerves.—Fl. June, July. Annual.

10. *T. glomeratum* (Smooth Round-headed Trefoil). —*Flowers* in round prickly heads, which are both terminal and axillary; *calyx-teeth* broad, very acute, re-

flexed ; *stems* prostrate.—Gravelly places near the sea ; not common. Very similar to *T. scabrum*, but the heads are rounder, and the calyx-teeth more leafy and spreading.—Fl. June. Annual.

11. *T. subterraneum* (Subterraneous Trefoil). — *Flowers* 3—5 together, in axillary heads, at first erect, in fruit abruptly bent down, and sending out branched fibres, which penetrate into the ground.—Dry banks ; not uncommon. A curious little plant, a few inches long, with prostrate branched stems and white flowers, which are remarkable for the above-named character of bending down and, by the help of the altered calyx, burying the seed in the ground while yet attached to the plant.—Fl. May, June. Annual.

* To this group belong several other species of Trefoil, which are of local occurrence, and therefore not likely to be met with by the beginner in botany. *T. stellatum* (Starry-headed Trefoil) occurs only near Shoreham, on the coast of Sussex. It is distinguished by the remarkably large calyx of the fruit, which spreads in a star-like manner. *T. Molinerii* (a variety of the commonly cultivated *T. incarnatum*), *T. Boccóni*, and *T. strictum*, were, until recently, considered to be confined to the southern shores of continental Europe, but are now known to be undoubtedly wild on the cliffs of the extreme south coast of Cornwall. The first is a short downy plant, with large soft heads of white flowers; *T. Boccóni* is a small plant, 2—4 inches high, with roundish heads of white or pink flowers, the heads always growing in pairs ; and *T. strictum*, which is of about the same size, has globular heads of whitish flowers, and is well distinguished by its 2-seeded legumes, which are bulged near the summit, and by its narrow-toothed leaflets, resembling in shape those of *Melilotus officinalis*. (*See* "A Week at the Lizard," p. 313.)

*** *Calyx inflated after flowering.*

12. *T. fragiferum* (Strawberry-headed Trefoil). — *Heads* globose, on long stalks ; *calyx,* after flowering, membranaceous, downy, and remarkably inflated ; *stem* creeping.—Meadows and pastures ; not very common. This plant has somewhat of the habit of *T. repens ;* but the flowers are light purple, and the heads of inflated calyces, which are often tinged with pink, are not unlike the fruit from which the plant receives its name. —Fl. July, August. Perennial.

**** *Standard withering but not falling off, finally bending down and covering the legume ; flowers yellow.*

13. *T. procumbens* (Hop Trefoil).—*Flowers* in dense, roundish, oblong heads ; *leaves* stalked.—Dry pastures ; abundant. Not unlike *Medicágo lupulína* in habit, but at once distinguished when in fruit by the hop-like heads of withered flowers. Specimens occur on the sea coast, 2—3 inches high, with scanty foliage, and comparatively large flowers.—Fl. June, July. Annual.

14. *T. filiformé* (Lesser Yellow Trefoil).—*Flowers* in loose heads, 3—12 together ; *leaves* scarcely stalked.— Dry pastures ; abundant. Like the last, but smaller. —Fl. June, July. Annual.

8. Lotus (*Bird's-foot Trefoil*).

1. *L. major* (Greater Bird's-foot Trefoil).—*Flowers* in umbels, 8—10 together ; *calyx-teeth* spreading like a star in the bud ; *stems* nearly erect, tubular.—In damp bushy places ; common. Whole plant more or less covered with soft hair ; 1—3 feet high ; stem weak, and usually supported by the bushes and other plants among which it grows.—Fl. deep yellow, July, August. Perennial.

2. *L. corniculatus* (Common Bird's-foot Trefoil).— *Flowers* in umbels, 8—10 together ; *calyx-teeth* straight

in the bud ; *stems* prostrate, not tubular.—Pastures ; abundant. A pretty flower, familiarly known among children by the name of " Shoes-and-stockings." The foliage is usually smooth, with a few scattered hairs, or

LOTUS CORNICULATUS (*Common Bird's-foot Trefoil*).

more rarely covered with long soft hairs. The flowers on the same plant, and even in the same umbel, vary from bright yellow to deep brownish orange.—Fl. July, August. Perennial.

* A third species, *L. angustissimus*, which grows in the west of England, is distinguished by its bearing heads of 3—4 small flowers ; the calyx-leaves are very long, and the whole plant is clothed with soft hairs.

ANTHYLLIS VULNERARIA (*Common Lady's Fingers*).

9. ANTHYLLIS (*Lady's Fingers*).

1. *A. Vulneraria* (Common Lady's Fingers, or Kidney Vetch).—The only British species. A handsome herbaceous plant, with pinnate *leaves* of a glaucous hue, (the terminal leaflet largest,) and yellow *flowers*, with

pale inflated *calyces*. The dense heads of flowers grow two together at the end of each stalk. On some of the sea-cliffs of Cornwall, and a few other places, the plant has a stunted habit of growth, and bears crimson, purple, cream-coloured, or white flowers.—Fl. June, August. Perennial.

10. Oxýtropis.

1. *O. Uralensis* (Hairy Mountain Oxytropis).—*Leaves* and *flowers* rising directly from the roots ; *flower-stalks* longer than the leaves, silky, like the rest of the plant ; *legume* 2-celled.—Dry mountain pastures in Scotland. — Fl. in heads, bright purple, June, July. Perennial.
  * Another species, *O. campestris,* (Yellowish Mountain Oxytropis,) grows in the Clova mountains. In this the *leaves* and *flower-stalks* are about equal in length ; the *flowers* are yellowish, tinged with purple, and the *legume* is imperfectly 2-celled.

11. Astrágalus (*Milk Vetch*).

1. *A. glycyphyllus* (Sweet Milk Vetch).—*Stem* prostrate ; *leaves* longer than the flower-stalks ; *legumes* curved, smooth.—Thickets, on a chalky or gravelly soil. Stem 2—3 feet long ; flowers in short dense spikes, dull yellow.—Fl. June, July. Perennial.
  2. *A. hypoglottis* (Purple Milk Vetch).—*Stem* prostrate ; *flower-stalks* longer than the leaves ; *legumes* erect, hairy.—Chalky and gravelly places. Well distinguished from the last by its slender stems, a few inches long, and by its purplish, sometimes white, flowers.—Fl. July. Perennial.
  * *A. alpínus* grows only in the glen of the Dole, Clova, and is distinguished by its spreading, branched *stem ;* spikes of drooping *flowers,* which are white, tipped with purple ; and by its pendulous *legumes,* clothed with black hairs.

VICIA CRACCA (*Tufted Vetch*).

12. Vícia (*Vetch*).

* Flower-stalks longer than the leaves ; many-flowered.

1. *V. sylvática* (Wood Vetch).—*Leaflets* in about 8 pairs, elliptical, abrupt, with a short point ; *stipules* crescent-shaped, deeply toothed at the base.—Mountainous woods, not common. A large and beautiful species, with a long stem, 3—6 feet high, climbing by

means of its branched tendrils. Flowers numerous, cream-coloured, with bluish veins.—Fl. July, August. Perennial.

2. *V. Cracca* (Tufted Vetch).—*Leaflets* in about 10 pairs, narrow, pointed, silky; *stipules* half - arrow shaped, scarcely toothed; *flowers* crowded in one-sided spikes.—Bushy places, frequent. One of the most ornamental of British plants, climbing along the tops of hedges, and adorning them with its slender spikes of blue and purple flowers.—Fl. July, August. Perennial.

** *Flowers axillary, scarcely stalked.*

3. *V. sativa* (Common Vetch).—*Flowers* in pairs, with very short stalks; *leaflets* oblong, in 5—7 pairs, the lower ones often inversely heart-shaped; *stipules* half-arrow shaped, toothed at the base, marked with a sunken dark spot; *seeds* smooth.—Fields; common. This species being extensively cultivated as fodder for cattle, varies considerably in luxurance, according to soil. It usually grows about 2 feet high, and bears blue and purple or red flowers. A small variety, *V. angustifolia* (Narrow-leaved Vetch), by some botanists considered a distinct species, has very narrow *leaves*, and crimson *flowers*.—Fl. June, July. Annual.

4. *V. lathyroídes* (Spring Vetch).—*Flowers* solitary, sessile; *legumes* smooth; *leaflets* in 2—3 pairs; *stipules* entire, not marked with a dark spot; *seeds* nearly cubical, roughish.—Dry places, but not very common. Nearly allied to the last, but very much smaller, rarely exceeding 6 inches in length.—Fl. bright purple, April, May. Annual.

5. *V. lútea* (Rough-podded Yellow Vetch).—*Flowers* solitary, sessile; *legumes* hairy; *stipules* marked with a deep red blotch.—Sea-coast; rare. About 2 feet high, with pale yellow, rather large flowers.—Fl. June, July. Perennial.

6. *V. sépium* (Bush Vetch).—*Flowers* in axillary

M

clusters of from 4—6 ; *legumes* smooth ; *leaflets* egg-shaped, obtuse, gradually decreasing in size towards the end of the leaf-stalk.—Woods and shady hedges ; common. Distinguished by its clusters of bluish purple flowers, which grow on short stalks in the axils of the leaves.—Fl. May, June.   Perennial.

* Other, but less common, species of Vetch, are *V. hýbrida* (*flowers* yellow), closely allied to *V. lutea,* but distinguished by having the outside of the *standard* clothed with abundance of shining yellowish hair ; this is found only on Glastonbury-Tor Hill, and at Swan Pool, near Lincoln : *V. lœvigata,* (*flowers* pale blue, or whitish,) allied to the last, but having smooth, not hairy, *legumes ;* growing in the pebbly shore of Weymouth, Dorsetshire : and *V. Bithýnica,* (*flowers* solitary, stalked, purple with whitish wings,) which has 4 pairs of *leaflets* on the lower *leaves,* and only 2 on the upper.   This last grows in bushy places, on a gravelly soil, near the sea.

13. Ervum (*Tare*).

1. *E. hirsútum* (Hairy Tare).—*Flowers* about 6 together ; *legumes* hairy, 2-seeded.—Fields and hedges ; very common.   A slender, much-branched plant, with the habit of a *Vetch,* forming tangled masses of stems and leaves, with minute bluish-white flowers.   This, though a mischievous weed, is not the Tare of the Holy Scriptures, which is supposed to be the *Darnel* (*Lolium temulentum*).—Fl. June, July.   Annual.

2. *E. tetraspermum* (Smooth Tare).—*Flowers* about 2 together, on a slender stalk ; *legumes* smooth, 4-seeded.—In similar situations with the last, but less common.   The flowers are light purple, and the whole plant much slenderer and less branched than *E. hirsutum.*—Fl. June, July.   Annual.

* *E. grácilé* (Slender Tare) is by some botanists con-

sidered a distinct species; others make it a variety of
the last. The *flowers* grow 1—4 together, and the

ERVUM HIRSUTUM (*Hairy Tare*).

legumes are 6—8-seeded. It has been found in Kent
and the Isle of Wight.

14. LÁTHYRUS (*Vetchling*).

* *Stalks many-flowered.*

1. *L. pratensis* (Meadow Vetchling).—*Leaf* of two
narrow leaflets ; *stipules* arrow-shaped, as large as the
leaflets.—Grassy places; common. A climbing plant,
2—3 feet long, with showy bright yellow flowers, which
all turn one way.—Fl. July, August. Perennial.

2. *L. sylvestris* (Narrow-leaved Everlasting Pea).—
Leaf of two sword-shaped leaflets ; *stem* winged.—
Woods and thickets ; not very common. Much resem-
bling the " Everlasting Pea " of Gardens (*L. latifolius*).
The stems climb to the height of 5—6 feet ; the flowers
are large, greenish yellow tinged with purple, but not

so handsome as the garden species.—Fl. July, August. Perennial.

LATHYRUS PRATENSIS (*Meadow Vetchling*).

3. *L. palustris* (Blue Marsh Vetchling).—*Leaf* of
2—4 pairs of very narrow acute leaflets; *stem* winged.—
Boggy meadows; rare. A climbing plant, with bluish
purple flowers, smaller than the last.—Fl. July, August.
Perennial.

4. *L. maritimus* (Sea-side Everlasting Pea).—*Flower-stalks* shorter than the leaves; *leaves* of 3—8 pairs
of egg-shaped leaflets; *stem* angular, not winged.—
Pebbly sea-shores; rare. Stems prostrate, about a span
long; flowers purple, variegated with crimson and
blue.—Fl. July, August. Perennial.

** *Stalks single-flowered.*

5. *L. Áphaca* (Yellow Vetchling).—*Tendrils* without
leaves; *stipules* very large, leaf-like, heart-arrow-shaped.
—Sandy and gravelly fields; rare. A pretty little plant,
with yellow flowers, and remarkable for being entirely
destitute of leaves, the place of which is supplied by a
pair of stipules at the base of each tendril.—Fl. June—
August. Annual.

6. *L. Nissólia* (Crimson Vetchling).—*Leaves* simple,
very narrow, destitute of tendrils.—Grassy places; not
common. A beautiful and well marked species, with
upright stems about a foot high, small bright crimson
flowers, and narrow, grass-like foliage.—Fl. June, July.
Annual.

* *L. hirsútus* (Rough-podded Vetchling), a very rare
species, has its *flowers* in pairs; pale blue, with a crimson standard, and hairy *legumes*.

15. Órobus *(Bitter Vetch).*

1. *O. tuberosus* (Tuberous Bitter Vetch).—*Leaves* of
2—4 pairs of oblong leaflets, which are glaucous beneath; *stipules* half-arrow-shaped; *stem* simple, erect,
winged.—Woods; not uncommon, especially in the west
of England. A pretty spring Vetch, with clusters of
blue and purple flowers in the axils of the leaves, grow-

ing in similar situations with the Wood Anemone, but appearing somewhat later. It may at once be distinguished from any of the true Vetches, by its being destitute of tendrils, the place of which is supplied by

ÓROBUS TUBEROSUS (*Tuberous Bitter Vetch*).

a soft bristle-like point. The roots are tuberous, and are "eaten by the Highlanders, under the name of *Cormeille*, a very small quantity being said to allay and prevent hunger." (*Sir W. J. Hooker.*)—Fl. May, June. Perennial.

    * Two other species of Órobus are natives of Britain; *O. niger* (Black Bitter Vetch), a rare Scottish plant, with very narrow *stipules*, and a branched *stem*, which is angular, but not winged ; and *O. sylvaticus* (Wood Bitter Vetch), also a northern plant, with 7—10 pairs of *leaflets*, and a branched, hairy *stem*.

ORNÍTHOPUS PERPUSILLUS (*Common Bird's-foot*).

16. ORNÍTHOPUS (*Bird's-foot*).

1. *O. perpusillus* (Common Bird's-foot).—The only British species.—Sandy heaths; common. A minute and very beautiful plant, with a spreading prostrate *stem*, downy *leaves* of 6—12 pairs of leaflets, and an odd one; heads of exceedingly small cream-coloured *flowers*, veined with crimson, with a leaf at the base of each head; and jointed *legumes*, which become curved as they ripen, and bear a singular resemblance to a bird's foot.—Fl. June—August. Perennial.

* *Arthrolóbium ebracteatum* (Sand Joint-Vetch), a little plant, in many respects similar to the last, has been found in the Scilly Islands. It may be distinguished by having no floral leaf.

HIPPOCRÉPIS COMÓSA (*Tufted Horse-shoe Vetch*).

17. HIPPOCRÉPIS (*Horse-shoe Vetch*).

1. *H. comósa* (Tufted Horse-shoe Vetch).—The only British species.—Common on chalky banks. A low

tufted plant, with much-branched *stems*, which are woody at the base, and elegant *leaves*, composed of 6—12 narrow leaflets. The umbels of yellow *flowers*

ONÓBRYCHIS SATÍVA (*Common Saint-foin*).

might be mistaken for those of *Lotus corniculatus*, but for the curious structure of the *seed-vessels*, which are shaped like a series of horse-shoes united by their extremities.—Fl. May—August. Perennial.

18. Onóbrychis (*Saint-foin*).

1. *O. sativa* (Common Saint-foin).—The only British species.—Chalky and limestone hills ; not uncommon. A handsome plant, often cultivated as fodder in dry, chalky and gravelly soils. The *stems* are 1—2 feet long ; the *leaves* are composed of 8—12 pairs of oblong leaflets, with an odd one, and the *flowers*, which grow in clusters, or rather spikes, are crimson, variegated with pink and white.—Fl. June, July.   Perennial.

Ord. XXVI.—ROSACEÆ.—The Rose Tribe.

Calyx most frequently 5-lobed, sometimes 4, 8, or 10-lobed ; *petals* 5, inserted on the calyx, regular ; *stamens* varying in number, generally more than 12, inserted on the calyx, curved inwards before the expansion of the petals ; *carpels* many, or solitary, either distinct, or combined with each other, and with the calyx ; *styles* distinct, often lateral ; *fruit* either a *drupe* (cherry, or plum),—an assemblage of erect capsules opening at the side,—a number of nut-like seeds inserted into a fleshy receptacle (Strawberry, Blackberry),— enclosed in the fleshy tube of the calyx (hip of the Rose), —or a *pome* (apple). —A large and important Order, containing about a thousand species, many of which, either in a wild or cultivated state, produce excellent fruit. Cherries, Plums, Almonds, Peaches, Nectarines, Apricots, Strawberries, Raspberries, Blackberries, Apples, Pears, and Quinces, all belong to this Order. It is to be noted, however, that valuable as these fruits are, the leaves, bark, flowers, and even seeds, of many, abound in a deadly poison, called hydrocyanic or prussic acid. The variety of form displayed by the fruit of the Rose tribe has afforded a facility for sub-dividing the Order into several Sub-orders, or Groups, the characters of which are subjoined.

Sub-order I.—AMYGDALEÆ.—*The Almond Group.*

In plants belonging to this division, the pistil is
solitary, and the fruit when ripe is a *drupe*, that is, a
single seed enclosed in a hard case, which is itself surrounded by a fleshy or juicy pulp, with an external
rind, or cuticle ; the bark often yields gum, and prussic
acid is generally abundant in the leaves and seeds.
They are shrubs or trees, and inhabit the cold and temperate regions of the northern hemisphere. Examples
of the deadly properties residing in these plants are
afforded by the leaves of the common Laurel, *Prunus*

Lauro-cérasus, even the vapour of which is destructive to insect life. The oil of Bitter Almonds is extremely poisonous, and many instances are recorded of its fatal effects. But notwithstanding the presence of this destructive principle in the leaves and other parts of the trees belonging to this division, the fruit is, with the exception of the Laurel, harmless, or even a nourishing food. *Amýgdalus communis*, the Almond-tree, grows naturally in Barbary, and in Asia, from Syria to Affghanistan, and is extensively cultivated in the south of Europe. There are two varieties of the tree, one yielding the sweet, the other the bitter Almond. Jordan Almonds, which are considered the best, are brought from Malaga ; bitter Almonds are imported from Mogadore. The varieties of *Amýgdalus Persica* produce the Peach, Nectarine, and Apricot. *Prunus domestica*, and its varieties, afford Plums of many kinds. *P. Lusitanica* is well known by the name of Portugal Laurel.—(*See* " Forest Trees," vol. i. pp. 237—274.)

1. PRUNUS (Plum and Cherry).—*Nut* of the *drupe* smooth, or slightly seamed.—(Name from the Greek *prouné*, a plum : *Cérasus*, a name sometimes given to one division of this genus, is derived from *Cerasus*, a city of Pontus, whence the Roman general Lucullus introduced a superior kind, B.C. 67.)

Sub-order II.—SPIRÆÆ.—*Meadow-sweet Group.*

This division contains a limited number of herbaceous or shrubby plants, which bear their seeds in dry, erect capsules, opening at the side, termed *follicles*. Several species of *Spiræa* are ornamental shrubs, and are commonly cultivated in gardens.

2. SPIRÆA (Meadow-sweet, Dropwort).—*Calyx* 5-cleft; *stamens* numerous ; *follicles* 3—12, bearing few seeds. (Name of Greek origin.)

Sub-order III.—Dryádeæ.—*The Strawberry Group.*

In this division the form of the fruit varies much more than in either of the preceding ; but in every case the calyx is permanent, and contains a number of nut-like seeds, with or without tails, placed on a pulpy, spongy, or dry receptacle ; in the Bramble, each seed is enveloped in pulp, the fruit being an assemblage of small drupes ; in Agrimony alone there are but two seeds, which are enclosed in a bristly, hardened calyx. The plants in this division are mostly herbaceous, but some few are shrubs. None of them are injurious ; the leaves and roots of some are astringent, or tonic. The fruit of the Strawberry, Raspberry, and Bramble, is too well known to need any description.

3. Dryas (Mountain Avens).—*Calyx* in 8—10 equal divisions, which are all in one row ; *petals* 8—10 ; *styles* finally becoming feathery tails, not hooked at the extremity. (Name from the Greek *drys*, an Oak, from a fancied resemblance between the leaves.)

4. Géum (Avens).—*Calyx* 10-cleft, in 2 rows, the outer divisions smaller ; *petals* 5 ; *styles* finally becoming jointed awns hooked at the extremity. (Name from the Greek *geyo*, to taste.)

5. Potentilla (Cinquefoil). — *Calyx* 10-cleft, in 2 rows, the outer divisions smaller ; *petals* 5 ; *seeds* without awns, on a dry receptacle. (Name from the Latin *potens*, powerful, from the powerful properties supposed to reside in some species.)

6. Tormentilla (Tormentil).—*Calyx* 8-cleft, in 2 rows, the outer divisions smaller ; *petals* 4 ; *seeds* without awns, on a dry receptacle. (Name from the Latin *tórmina*, dysentery, for which disease it was considered a specific.)

7. Sibbaldia.—*Calyx* 10-cleft, in 2 rows, the outer divisions smaller ; *petals* 5 ; *stamens* 5 ; *seeds* about 5, without awns, on a dry receptacle. (Named after Robert Sibbald, a Scottish naturalist of the 17th century.)

8. CÓMARUM (Marsh Cinquefoil).—*Calyx* 10-cleft, in 2 rows, the outer divisions smaller ; *petals* 5 ; *seeds* without awns, on an enlarged spongy receptacle. (From the Greek *cómaros*, the name of a plant very distinct from the present.)

9. FRAGARIA (Strawberry).—*Calyx* 10-cleft, in 2 rows, the outer divisions smaller ; *petals* 5 ; *seeds* without awns, on an enlarged, fleshy receptacle. (Name from the Latin *fragum*, a Strawberry, and that from *fragrans*, fragrant.)

10. RÚBUS (Bramble).—*Calyx* 5-cleft ; *petals* 5 ; *fruit* an assemblage of small drupes, arranged on and around a spongy receptacle. (Name from the Latin, *ruber*, red.)

11. AGRIMONIA (Agrimony). — *Calyx* 5-cleft, top-shaped, covered with hooked bristles ; *petals* 5 ; *stamens* about 15 ; *seeds* 2, enclosed in the tube of the hardened calyx. (Name of Greek origin.)

Sub-order IV.—SANGUISORBEÆ.—*The Burnet Group.*

The plants of this group would seem at first sight to be scarcely connected with those already described. It will, however, be found, on a close examination, that in many important respects they agree with the characters given in the description of the *Order* ROSACEÆ, though in others scarcely less important, they appear to differ ; these are the absence of petals, and the hardened calyx of the fruit containing 1 or 2 nut-like seeds. The calyx is 3—8 cleft, and the stamens are usually few in number. The plants are either herbaceous or shrubby, and like those of the last group, their properties are astringent, or tonic. In some species the flowers grow in round or oblong heads.

12. ALCHEMILLA (Lady's Mantle).—*Calyx* 8-cleft, in 2 rows, the outer divisions smaller ; *petals* 0 ; *stamens* 1—4, opposite the smaller divisions of the calyx ; *seeds* 1 or 2, enclosed in the hardened calyx. (Name from its pretended value in Alchemy.)

13. SANGUISORBA (Burnet).—*Calyx* 4-cleft, coloured, (not green,) with 2—4 scale-like bracts at the base; *petals* 0; *stamens* 4; *seeds* 1 or 2, enclosed in the tube of the hardened calyx. (Name from the Latin *sanguis*, blood, and *sorbeo*, to staunch, from the supposed virtues of the plant.)

14. POTÉRIUM (Burnet Saxifrage).—*Stamens* and *pistils* in separate flowers; *flowers* in heads; *calyx* 4-cleft, coloured, with 3 scale-like bracts at the base; *petals* 0; *stamens* numerous; *stigma* tufted. (Name from the Greek *potérion*, a drinking cup, the plant being used in the preparation of *Cool-tankard*.)

Sub-order V.—ROSEÆ.—*The Rose Group.*

This division contains the genus from which both the *Order* and *Sub-order* take their names. Here, also, the fruit furnishes the main characteristic; it consists of a number of nut-like hairy seeds, enclosed within the

fleshy tube of the calyx, which is contracted at the top. The Roses are shrubs more or less prickly, (not thorny,) with pinnate leaves. The number of species is very great, of varieties incalculable, the beauty and fragrance of the flowers having rendered them favourite objects of cultivation from a very early period. The most fragrant of the British Roses are the common Dog Rose, *Rosa canína*, and the Scotch Rose, *R. spinosissima*, the latter being the origin of the numerous varieties of double Scotch Rose. From the petals of *R. centifolia* and

R. Damascéna are made Rose-water, and Attar of Roses. It is stated that 100,000 Roses, the produce of 10,000 bushes, yield only nine drams of Attar. Some species, as *R. rubiginosa*, Sweet-brier, are copiously furnished with glands which secrete a fragrant viscid fluid. From the pulp of the fruit, called a hip, is made a conserve, which is used in the preparation of various medicines ; and the woody stems of the Dog-rose have of late years been much sought after for making walking-sticks.

15. ROSA (Rose).—*Calyx* urn-shaped, contracted at the mouth, and terminating in 5, often leaf-like, divisions ; *petals* 5 ; *stamens* numerous ; *seeds* numerous. (Name from the Latin, *rosa*, and that from the Greek, *rhodon*, its ancient names.)

Sub-order VI.—POMEÆ.—*The Apple Group.*

In the plants of this division the fruit is what is called a *pome;* that is, the tube of the calyx enlarges and becomes a fleshy or mealy fruit, enclosing 1—5 cells, which are either horny, as in the Apple, or bony, as in the Medlar. The Apple Group contains many well-known fruit-trees, namely the Apple, Pear, Quince, Medlar. Service, Mountain Ash, and Hawthorn. The seeds, and occasionally the flower and bark of some, yield prussic acid. All the cultivated varieties of Apple are derived from the wild Apple, or Crab, *Pyrus Malus;* the garden Pears from a thorny tree, with hard astringent fruit, *Pyrus commúnis.* The wood of the Pear is almost as hard as Box, for which it is sometimes substituted by wood-engravers. The fruit of the Mountain Ash, and some other species, yields malic acid, and the leaves prussic acid, in as great abundance as the Laurel. All the plants of this division are either trees or shrubs.

16. PYRUS (Pear, Apple, Service, and Mountain Ash). —*Calyx* 5-cleft ; *petals* 5 ; *styles* 2—5 ; *fruit* fleshy, or juicy, with 5 horny, 2-seeded cells. (Name from the Latin *pyrus*, a pear.)

17. MÉSPILUS (Medlar).—*Calyx* 5-cleft, divisions leaf-like; *petals* 5; *styles* 2—5; *fruit* fleshy, top-shaped, terminating abruptly, with the ends of the bony cells exposed. (Name from the Greek *mespilé*, a medlar.)

18. CRATÆGUS (Hawthorn).—*Calyx* 5-cleft, divisions acute; *petals* 5; *styles* 1—5; *fruit* oval, or round, con-cealing the ends of the bony cells. (Name from the Greek *cratos*, strength, in allusion to the hardness of the wood.)

PRUNUS SPINOSA (*Sloe, or Blackthorn*).

1. PRUNUS (*Plum and Cherry*).

* *Fruit covered with bloom; young leaves rolled together.*

1. *P. spinosa* (Sloe, or Blackthorn).—*Branches* very thorny; *leaves* narrow, elliptical, smooth above, slightly

N

downy near the mid-rib below ; *flowers* mostly solitary.—
Woods and hedges ; abundant. A well-known thorny
bush, which probably derived its name Blackthorn from
the hue of its bark, which is much darker than that of
the Hawthorn. The flowers appear in March and April,
and usually before the leaves have begun to expand.
The latter are used to adulterate tea. The fruit is small,
nearly round, and so austere, that a single drop of its
juice placed on the tongue will produce a roughness on
the throat and palate which is perceptible for a long
time. It enters largely into the composition of spurious
port wine.—Fl. March—May.   Shrub.

2. *P. insitítia* (Bullace). — *Branches* ending in a
thorn ; *leaves* elliptical, downy beneath ; *flowers* in pairs.
—Woods and hedges.   Larger than the last, and producing a more palatable fruit.   The leaves and flowers
expand about the same time.   This is by some botanists
considered merely a variety of the preceding.   *P. domés-
tica*, Wild Plum, appears to be as closely connected with
the Bullace as that is with the Sloe ; the branches are
thornless, and the fruit oblong.   From one or other of
these three all the cultivated varieties of Plum are supposed to have originated.—Fl. April, May.   Small tree.

** *Fruit without bloom ; young leaves folded together.*

3. *P. Padus* (Bird-Cherry).—*Flowers* in clusters ;
leaves narrow, egg-shaped ; *fruit* oblong.—A handsome
shrub, or small tree, not uncommon in the north of
England in a wild state, and very generally admitted
into gardens and shrubberies elsewhere.   The clusters
of flowers and drupes are not unlike those of the Portugal Laurel, but the leaves are not evergreen.—Fl.
white, May.   Small tree.

4. *P Avium* (Wild Cherry).—*Flowers* in umbels ;
leaves drooping, suddenly pointed, downy beneath ;
calyx-tube contracted above ; *fruit* heart-shaped. —
Woods and hedges ; common.   A highly ornamental

tree, not only on account of its elegant white flowers in spring, but even more so in autumn, when its leaves assume a bright crimson hue, which distinguishes them

PRUNUS PADUS (*Bird-Cherry*).

among all the varied tints of the fading year. The fruit is small, bitter, black, or red, and as soon as it is ripe is greedily devoured by birds.—Fl. May.   Lofty tree.

5. *P. Cérasus* (Red Cherry).—*Flowers* in umbels ; *leaves* not drooping, smooth on both sides ; *calyx-tube* not contracted ; *fruit* round.—Woods and hedges. This species is distinguished from the preceding by the cha-

racters given above, and by its lower stature, which is
said not to exceed 8 feet, while the other attains a height
of 30—40 feet. The fruit also differs in being juicy,
acid, and always red. Some botanists, however, consider
them mere varieties of the same tree. From one or
other all the cultivated kinds of Cherry are derived.—
Fl. May. Shrub.

2. SPIRÆA (*Meadow-sweet*).

1. *S. Ulmária* (Meadow-sweet, Queen of the Meadows).—Herbaceous ; *leaves* pinnate, the alternate leaflets smaller, white with down beneath, terminal leaflet
very large and lobed ; *flowers* in compound erect cymes.
Moist meadows ; common. A tall plant, 2—4 feet
high, with densely crowded yellowish white flowers,
which are elegant and fragrant.—Fl. July, August.
Perennial.

2. *S. Filipéndula* (Dropwort).—Herbaceous ; *leaves*
pinnate, with the alternate leaflets smaller, all deeply
cut into narrow, serrated segments ; *flowers* in a panicled cyme.—Dry pastures, especially on a limestone soil,
about a foot high ; well distinguished from the last by
its elegantly cut foliage, and less crowded flowers, the
petals of which are pink externally before they expand,
and when open are white and scentless. A variety with
double flowers is common in gardens.—Fl. July—September. Perennial.

* *S. salícifolia* (Willow-leaved Spiræa), is a shrubby
species, with spike-like clusters of rose-coloured flowers,
and simple (not pinnated) leaves. It is occasionally
found in Scotland and the north of England, but is not
considered to be a native. Several other species of
Spiræa are common in gardens, to which they are very
ornamental. They are easily propagated by cuttings,
and will grow in any moist situation.

SPIRÆA ULMARIA (*Meadow-sweet, Queen of the Meadows*).

DRYAS OCTOPÉTALA (*Mountain Avens*).

3. DRYAS (*Mountain Avens*).

D. octopétala (Mountain Avens).—The only British species; not uncommon in the mountainous parts of England, Scotland, and Ireland, and at once distinguished from all other plants of the Order by its oblong deeply-cut leaves, which are white with woolly down beneath, and by its large and handsome white flowers, each of which has 8 petals.—Fl. June, July. Perennial.

GEUM URBANUM (*Common Avens, Herb Bennet*).

4. GÉUM (*Avens*).

1. *G. urbánum* (Common Avens, Herb Bennet).—
Flowers erect ; *awns* rigid ; *root-leaves* pinnate, with
smaller leaflets at the base ; *stem-leaves* ternate.—Hedges
and thickets ; common. A somewhat slender, little-
branched plant, 1—2 feet high, with yellow flowers,
which are less conspicuous than the round heads of
awned seeds which succeed them ; the stipules are large,
rounded, and cut.—Fl. June—August. Perennial.

2. *G. riválé* (Water Avens). — *Flowers* drooping ;

awns feathery; *root-leaves* pinnate, with the alternate leaflets and those at the base smaller; *stem-leaves* ternate.—Wet, mountainous woods; not unfrequent. Much stouter than the last, and well distinguished by the above characters, as well as by its larger flowers, of which the calyx is deeply tinged with purple, and the petals are of a dull purplish hue, with darker veins.— Fl. June, July. Perennial.

* A variety, *G. intermedium*, is sometimes found, which appears to partake of the characters of both the above species.

5. POTENTILLA (*Cinquefoil*).

* *Leaves pinnate.*

1. *P. anserína* (Silver-weed, Goose-grass). — *Leaves* pinnate, the alternate leaflets smaller; *leaflets* sharply cut, silky on both sides, especially beneath; *flower-stalks* solitary, axillary. — Waste ground; common. Well marked by its creeping stem, elegantly cut, silky foliage, and showy yellow flowers.—Fl. June, July. Perennial.

* To the division with pinnate leaves belong *P. fruti-cosa* (Shrubby Cinquefoil), a bushy species, 3—4 feet high, with hairy leaves and large yellow flowers, which last grow several together at the end of the stems: bushy places; rare: and the yet more uncommon species, *P. rupestris* (Rock Cinquefoil), which has large white flowers, and is found only in Montgomeryshire.

** *Leaflets 5 on a stalk (quinate).*

2. *P. reptans* (Creeping Cinquefoil).—*Stem* creeping, rooting at the joints; *leaves* stalked; *leaflets* inversely egg-shaped, tapering at the base, serrated; *flower-stalks* solitary.—Meadows and way-sides; common. *Flowers* handsome, yellow, on long stalks.—Fl. June—August. Perennial.

3. *P. argéntea* (Hoary Cinquefoil).—*Stem* prostrate;

leaflets inversely egg-shaped, cut, white and downy beneath, their edges rolled back.—Pastures and road-sides, on a gravelly soil ; not common. *Flowers* yellow, small, several together at the ends of the stems.—Fl. June. Perennial.

POTENTILLA ANSERÍNA *and* P. REPTANS.

4. *P. verna* (Spring Cinquefoil).—*Stem* prostrate ; *leaflets* sometimes 7 on the root-leaves, inversely egg-shaped, serrated towards the end, hairy on the edge and ribs beneath, not downy.—Dry pastures in various parts of England, but not common. A small woody plant, about 5 inches long, with yellow flowers, 2—3 together at the ends of the stems.

* *P. alpestris* (Alpine Cinquefoil), is considered by some botanists a variety of the preceding ; by others, a distinct species. It is somewhat larger, and the stem has a more upright growth. *P. opáca* (Saw-leaved, Hairy Cinquefoil) occurs only on the hills of Clova and Braes of Balquidder, Scotland. The leaves are deeply cut throughout, and the stem is stouter than in the two last, but in other respects it is very like.

*** *Leaves 3 on a stalk, ternate.*

5. *P. Fragariastrum* (Strawberry-leaved Cinquefoil). —*Stem* prostrate ; *leaflets* inversely egg-shaped, cut, silky on both sides ; *petals* equalling the calyx in length. —Banks and hedges ; abundant. One of the earliest spring flowers, often confounded by young botanists with the wild Strawberry, *Fragaria vesca.* It may, however, be always distinguished by its prostrate mode of growth, and short, notched petals; the flower-stalks of the Strawberry being erect, and the petals entire.—Fl. January— May. Perennial.

* *P. tridentáta* (Three-toothed Cinquefoil), a rare species, found only on the Clova hills, is distinguished from the last by its oblong, wedge-shaped leaflets, each of which ends in 3 points. The petals are white, and longer than the calyx.

Upwards of a hundred and fifty species of *Potentilla* are described by botanists, inhabiting both the Eastern and Western Hemispheres, and preferring, generally, moist or rocky situations. They are all easy of cultivation, and some of them are handsome when in flower. They will grow in any common garden soil, and are easily increased, either by dividing the plants or by seed.

TORMENTILLA OFFICINALIS (*Common Tormentil*).

6. TORMENTILLA (*Tormentil*).

1. *T. officinalis* (Common Tormentil). — *Leaves* of 3 leaflets, ternate, sessile ; *root-leaves* of 5 leaflets (quinate) stalked ; *leaflets* narrow, acute, cut ; *stem* ascending.—Banks and woods ; common. Closely allied to the preceding genus, from which it is distinguished by its having only 4 petals ; but even this character is not constant. A small plant, with bright yellow flowers, and very woody roots, which latter are astringent, and are employed in medicine. Specimens are not uncommon in which the stem is prostrate ; this is by some botanists considered a distinct species, and is called

T. reptans (Trailing Tormentil).—Fl. all the summer. Perennial.

SIBBALDIA PROCUMBENS (*Procumbent Sibbaldia*).

7. SIBBALDIA.

1. *S. procumbens* (Procumbent Sibbaldia).—The only British species.—A small prostrate plant, with ternate, hairy leaves, and yellowish flowers, growing abundantly on the summits of the Highland mountains. The leaflets are wedge-shaped, and end in 3 points. The number of stamens and pistils is very variable.—Fl. June, July. Perennial.

8. CÓMARUM (*Marsh Cinquefoil*).

1. *C. palustré* (Marsh Cinquefoil).—The only species. —A herbaceous bog-plant, growing about a foot high,

COMARUM PALUSTRE (*Marsh Cinquefoil*).

with much of the habit of a Potentilla, but stouter. The lower *leaves* are usually of 7 long, cut leaflets, the upper of 5 or 3; and each *stem* bears several *leaves*, and a number of large dingy purple *flowers*.—Fl. July. Perennial.

FRAGARIA VESCA (*Wood Strawberry*).

9. FRAGARIA (*Strawberry*).

1. *F. vesca* (Wood Strawberry).—*Calyx* of the fruit bent back ; *hairs* on the general flower-stalk widely spreading, on the partial flower-stalks close-pressed, silky.—Woods and thickets ; common. A well-known plant, with bright-green hairy leaves, rooting stems, and erect flower-stalks. By these last two characters, as well as by the drooping fruit, this plant may be distin·guished from *Potentilla Fragariastrum* (Strawberry-leaved Cinquefoil), which is often mistaken for it by young botanists. The Strawberry derives its name either from the custom of laying straw between the rows of plants in gardens, or from the habit adopted by children of stringing the fruit on straws of grass.—Fl. May—July. Perennial.

* *F. elatior* (Hautboy Strawberry), which is occasionally found in situations where it is apparently, but not really, wild, has its flower-stalks clothed throughout with spreading hairs. It is also larger and more hairy than the last. Some plants of this last species bear flowers with stamens only, and consequently produce no fruit. This is also the case with some garden varieties, a fact which should be attended to by growers of Strawberries.

10. RUBUS (*Bramble*).

1. *R. Idæus* (Raspberry).—*Stem* nearly erect, round, downy and prickly ; *leaves* pinnate, of 5 leaflets, which are white, and very downy beneath ; *flowers* drooping ; *fruit* downy.—Rocky woods ; not very common. The origin of all the varieties of garden Raspberry, from which it differs principally in the size of the fruit, which is scarlet, and of an agreeable flavour.—Fl. May, June. Shrub.

2. *R. fruticósus* (Common Bramble, or Blackberry). —*Stem* arched, angular, prickly, rooting ; *leaves* of 5 crowded leaflets ; *flowers* erect, in compound panicles ; *calyx* of the fruit spreading, or bent back.—Common everywhere. This description includes a large number of species and varieties to which names have been severally given, but it is not here thought necessary to describe the characters at length, the genus being confessedly a difficult one, and on other accounts uninviting to the young botanist.—Most of the species flower from July to August, and ripen their fruit in September and October. Shrub.

3. *R. cæsius* (Dewberry). — *Stem* prostrate, nearly round, prickly below, bristly above ; *leaves* of 3 or 5 leaflets ; *panicle* simple ; *calyx* clasping the fruit.— Thickets and borders of fields ; not uncommon. In this species the fruit, which consists of a few large drupes,

is half enclosed in the calyx, and is covered with a grey
bloom.—Fl. June—August.   Shrub.

RUBUS FRUTICÓSUS (*Blackberry*).

4. *R. saxátilis* (Stone Bramble).—*Stem* herbaceous,
rooting ; *prickles* few, small ; *leaves* of 3 leaflets ; *flowers*
few together.—Stony, mountainous places, especially in
the north.   A small herbaceous species, about a span
high, with greenish-yellow flowers, and scarlet fruit,
consisting of 1—4 large drupes.—Fl. July, August.
Perennial.

5. *R. Chamæmórus* (Cloudberry).—*Stem* herbaceous, without prickles, 1-flowered ; *leaves* simple, lobed.—A very distinct species, growing in the mountainous parts of Great Britain and Ireland. About a span high, with large white or rose-coloured flowers, which are solitary, and bear the stamens and pistils on separate plants. The fruit is orange-red, and of a pleasant flavour.—Fl. June. Perennial.

11. AGRIMONIA (*Agrimony*).

1. *A. Eupatória* (Common Agrimony).—The only British species.—Common in waste ground. A slender herbaceous plant, 1—2 feet high, very different in habit from any of the preceding. The *leaves* are pinnate, with the alternate leaflets smaller, and all are deeply cut. The *flowers* are yellow, and grow in long tapering *spikes*. The whole plant is covered with soft hairs, and when bruised emits a slightly aromatic scent. Its properties are said to be tonic, and on this account it is often collected by village herbalists, and made into tea.—Fl. July, Aug. Perennial.

12. ALCHEMILLA (*Lady's Mantle*).

1. *A. vulgáris* (Common Lady's Mantle). —*Leaves* kidney-shaped, plaited, 7—9 lobed, lobes blunt, serrated; *flowers* in loose, divided clusters.—Hilly pastures; not uncommon. A herbaceous plant, about a foot high, with large and handsome soft leaves, and numerous small, yellowish-green flowers. — Fl. June — August. Perennial.

2. *A. alpína* (Alpine Lady's Mantle).—*Leaf* of 5—7 oblong, blunt leaflets, serrated at the end, white and satiny beneath.—Mountains in Scotland and the north of England. A very beautiful plant, remarkable for the lustrous, almost metallic hue of the under side of its leaves.—Fl. July, August. Perennial.

AGRIMONIA EUPATORIA (*Common Agrimony*).

ALCHEMILLA ARVENSIS (*Field Lady's Mantle*).

3. *A. arvensis* (Field Lady's Mantle, or Parsley Piert).
Leaves 3-cleft, wedge-shaped, downy ; *lobes* deeply cut ;
flowers tufted, sessile in the axils of the leaves.—Common everywhere. A small, inconspicuous weed, 3—6
inches long, with minute greenish flowers, which are
almost concealed by the leaves, and their large stipules.
Fl. May—August. Annual.

13. Sanguisorba (*Burnet*).

1. *S. officinalis* (Common Burnet).—The only British
species.—A tall and not inelegant plant, with pinnate
leaves ; erect, branched *stems*, sparely clothed with

leaves; and oblong heads of deep purple-brown *flowers.*
Moist pastures; not uncommon.—Fl. June—September.
Perennial.

SANGUISORBA OFFICINALIS (*Common Burnet*).

POTERIUM SANGUISORBA (*Salad Burnet*).

14. POTERIUM (*Salad Burnet*).

1. *P. sanguisorba* (Salad Burnet).—The only British species.—Common in dry pastures, especially on chalk and limestone. Not unlike the last in habit, but much smaller. The *leaves* are pinnate, with serrate leaflets, and have the taste and smell of cucumber. The *flowers* grow in small round heads, and are remarkable for their greenish-purple hue. The upper flowers in each head bear crimson-tufted *pistils*, the lower ones 30—40 *stamens*, with very long drooping filaments.—Fl. July, August. Perennial.

15. Rosa (*Rose*).

* *Shoots thickly set with bristles, mixed with nearly straight prickles.*

1. *R. spinosissima* (Burnet-leaved Rose).—*Leaflets* small, simply serrated, smooth ; *calyx* simple ; *fruit* nearly round.—Waste places, especially near the sea. A thick, very prickly bush, 2—4 feet high, with small, elegant foliage, and large cream-coloured flowers, which are deliciously fragrant ; *fruit* dark purple or black. The origin of the garden varieties of Scotch Rose.— May, June.—Shrub.

** *Shoots without bristles ; prickles straight, or slightly curved.*

2. *R. tomentósa* (Downy - leaved Rose). — *Leaflets* doubly serrated, downy, and glandular ; *calyx* pinnate.— Hedges and thickets ; not very common. Distinguished by its stout and long *shoots ;* downy, almost hoary, *leaves ;* large, deep red *flowers,* and very long *seed-vessels,* which are usually crowned with the copiously pinnate *calyx leaves.*—Fl. June, July. Shrub.

*** *Prickles, some hooked, some straight, mixed with bristles.*

3. *R. rubiginosa* (Sweet Brier).—*Leaflets* doubly serrated, hairy, glandular beneath, mostly rounded at the base ; *calyx* pinnate, remaining attached to the ripe fruit ; *fruit* pear-shaped when young.—Bushy places, especially on chalk. The Eglantine of the poets, but not of Milton ; the "twisted Eglantine" of that author being the Woodbine or Honeysuckle. A favourite garden plant, deservedly cultivated for the sake of its deliciously fragrant foliage.—Fl. June, July. Shrub.

**** *Bristles none ; prickles hooked.*

4. *R. canina* (Dog Rose).—*Leaves* smooth, or slightly hairy ; *calyx* pinnate, not remaining attached to the fruit ; *styles* distinct. — Hedges and bushy places ; abundant. This is the Common Hedge Rose, a flower

belonging exclusively to summer, and welcomed at its first appearance scarcely less warmly than the early Primrose in spring. The colour of the flower varies from white to a deep blush, and the leaves also differ considerably; but the above characters will be found to include all the principal varieties.—Fl. June, July. Shrub.

ROSA CANINA (*Dog Rose*).

5. *R. arvensis* (Trailing Dog Rose).—*Prickles* on the young shoots feeble; *leaves* smooth; *calyx* slightly pinnate, not remaining attached to the fruit; *styles* united; *stigmas* forming a round head.—Woods and hedges; common in the south of England. Distinguished from all the other British species of Rose by its slender, trailing stems. The *flowers* are white and

scentless, and there are fewer prickles than in most
other species.—Fl. June—August.   Shrub.

* Botanists describe no less than nineteen species of
native Roses, but, as many of these are rare, and the
characters of others are difficult to discriminate, it has
been thought best to describe here only those which
are of common occurrence, or otherwise remarkable.

PYRUS COMMUNIS (*Wild Pear*).

16. PYRUS (*Pear, Apple, Service, and Mountain Ash*).

1. *P. commúnis* (Wild Pear).—*Leaves* simple, egg-
shaped, serrated ; *flowers* in corymbs ; *fruit* tapering at

the base.—Woods and hedges. A small upright tree, often bearing thorns at the extremities of its branches. The seed-vessel, in a wild state, is woody, austere, and worthless, yet is converted by cultivation into a most luscious fruit.—Fl. white, April, May. Tree.

PYRUS MALUS (*Crab Apple*).

2. *P. Malus* (Crab Apple).—*Leaves* simple, egg-shaped, serrated; *flowers* in a sessile umbel; *styles* combined below; *fruit* with a hollow beneath.—Woods and hedges. A small spreading tree, with thornless branches,

umbels of white flowers delicately shaded with pink, and nearly round fruit, which is intensely acid. It was formerly much used in making verjuice, and in the preparation of pomatum, so called from *pomum*, an apple.— Fl. May. Tree.

PYRUS TORMINALIS (*Wild Service Tree*).

3. *P. torminális* (Wild Service Tree).—*Leaves* egg-shaped, with several deep, sharp lobes; *flowers* in corymbs.—Woods and hedges in the south of England. A small tree, with leaves shaped somewhat like those of the Hawthorn, but larger, and white flowers, which are succeeded by brownish, spotted, berry-like fruit.— Fl. May. Tree.

PYRUS AUCUPARIA (*Fowlers' Service Tree*).

4. *P. aucuparia* (Fowlers' Service, Mountain Ash, Quicken, or Rowan Tree).—*Leaves* pinnate, serrated; *flowers* in corymbs; *fruit* nearly round.—Mountainous woods. One of the most elegant of British trees, conspicuous in the flowering season by its delicate green foliage, large *corymbs* of blossom, and in autumn by its clusters of scarlet berry-like *pomes*, which are greedily eaten by birds, and often used as a lure by the bird-catcher or fowler, *auceps*. For a further account of

this tree, as well as the rest of the genus, see " Forest Trees of Britain," vol. i.—Fl. May.　Tree.

PYRUS ARIA (*White Beam Tree*).

5. *P. Aria* (White Beam Tree).—*Leaves* egg-shaped, cut, very white below ; *flowers* in corymbs ; *fruit* nearly round.—Woods.　A small tree, well distinguished by its very large leaves, which are remarkably white and silky beneath, especially when first expanding.　Nowhere is this tree more ornamental than on the ruinous walls of the ancient Roman town of Silchester, where it abounds.—Fl. June.　Tree.

17. Méspilus (*Medlar*).

1. *M. Germánica* (Common Medlar).—A tree well known in a cultivated state, but nowhere, perhaps, in Great Britain truly wild.　The *flowers* are white and

very large, and the *fruit* is remarkably flattened at the top, exposing the upper ends of the long seed-cells.—Fl. May. Tree.

CRATÆGUS OXYACANTHA (*Hawthorn*).

18. CRATÆGUS (*Hawthorn*).

1. *C. oxyacantha* (Hawthorn, White-thorn, or May-bush).—A tree which, though it varies considerably in its mode of growth, shape of its leaves, and colour of its flowers and fruit, is so well known as to need no description. The name Hawthorn is supposed to be a corruption of the Dutch *hæg*, or *hedge* thorn; although, therefore, the fruit is generally called a *haw*, that name is derived from the tree which produces it, and the tree does not, as is frequently supposed, take its name from the fruit which it bears.—Fl. May, and, in the mountains, till late in June. Tree.

Ord. XXVII.—ONAGRARIÆ.—The Willow-herb Tribe.

Calyx of 4, sometimes 2 lobes, which in the bud are attached to each other by their edges ; *calyx-tube* more or less united to the ovary ; *petals* as many as the lobes of the calyx, twisted while in bud ; *stamens* 4 or 8, rarely 2, springing from the mouth of the calyx ; *ovary* of 2 or 4 cells, often crowned by a disk ; *stigma* knobbed, or 4-lobed ; *fruit* a berry, or 4-celled capsule. —Herbaceous plants or shrubs, principally inhabiting the temperate parts of the globe, especially America and Europe. In this Order we find the elegant American genus *Fuchsia*, with its coloured 4-cleft calyx, and often edible fruit. Many species of *Œnothéra* are commonly cultivated as garden plants, some bearing flowers 3 or 4 inches in diameter ; those with yellow or white flowers which open only in the evening, are called Evening Primroses. The properties of the plants which compose this Order are unimportant. The wood of the *Fuchsia* is said to be used as a dye, and the roots of *Œnothéra biennis*, the common Evening Primrose, are eatable. In all, the number 4 predominates.

1. Epilóbium (Willow-Herb).—*Calyx* 4-parted, the lobes not combined after expansion ; *petals* 4 ; *stamens* 8 ; *capsule* long, 4-sided, 4-celled, 4-valved ; *seeds* numerous, tufted with down. (Name from the Greek *epi*, upon, and *lobos*, a pod, the flowers being placed on the top of a pod-like seed-vessel.)

2. Œnothéra (Evening Primrose).—*Calyx* 4-parted, the lobes more or less combined after expansion, and bent back ; *stamens* 8 ; *capsule* 4-celled, 4-valved ; *seeds* numerous, not bearded. (Name in Greek signifying " catching the flavour of wine.")

3. Isnardia.—*Calyx* 4-parted ; *petals* 4, or none ; *stamens* 4 ; *capsule* inversely egg-shaped, 4-angled, 4-celled. 4-valved, crowned with the calyx. (Named

after a French botanist of the 18th century, Antoine
d'Isnard.)

4. CIRCÆA (Enchanter's Nightshade).—*Calyx* 2-parted;
petals 2 ; *stamens* 2 ; *capsule* 2-celled, each cell containing a seed. (Name from Circe, the enchantress so celebrated in Greek Mythology.)

1. EPILÓBIUM (*Willow-Herb*).

* *Petals unequal in size ; stamens bent down.*

1. *E. angustifolium* (Rose Bay, or Flowering Willow).
—*Leaves* narrow, pointed, smooth.—Damp woods ; rare.
A tall and handsome species not often met with in a
wild state, but very common in gardens, where it is cultivated for the sake of its large, rose-coloured flowers.
Caution should be used in admitting it into a small
garden, as its roots creep extensively, and are very difficult to eradicate —Fl. July.   Perennial.

* *Petals all equal ; stamens erect ; stigmas 4-cleft.*

2. *E. hirsútum* (Great Hairy Willow Herb).—Woolly ;
leaves clasping the stem, narrow-oblong, serrated ; *stem*
much branched ; *root* creeping.—Wet places ; common.
A handsome species, 4—6 feet high, with large rose-
coloured flowers.   Well marked by its very downy stems
and leaves, and creeping roots.—Fl. July, August.
Perennial.

3. *E. parviflórum* (Small-flowered, Hairy Willow-
Herb).—Downy ; *leaves* sessile, narrow, toothed ; *stem*
nearly simple ; *root* fibrous.—Wet places ; common.
Distinguished from the last by its smaller size, un-
branched mode of growth, and fibrous roots.—Fl. July,
August.   Perennial.

4. *E. montánum* (Broad, Smooth-leaved Willow-Herb).
—*Leaves* egg-shaped, acute, smooth, toothed, the lower
ones shortly stalked ; *stem* round, slightly downy.—Dry
banks, and hilly places ; frequent.   A small species,

about a foot high, with rose-coloured flowers, which are most frequently found in a half-open state. It may often be detected when in seed by its capsules, the valves of which open lengthwise, and disclose the numerous *seeds* bearded with cottony down.—Fl. July, August. Perennial.

EPILOBIUM ANGUSTIFOLIUM (*Rose Bay, or Flowering Willow*).

*** *Petals all equal ; stamens erect ; stigma knobbed, not 4-cleft.*

5. *E. tetragónum* (Square-stalked Willow-Herb).— *Leaves* narrow, sessile, toothed ; *stem* 4-angled, nearly smooth.—Wet places ; common.   From 1 to 2 feet high, and often, like the last and following species, more conspicuous when in seed than in flower.—Fl. July, August. Perennial.

6. *E. palustré* (Narrow-leaved Marsh Willow-Herb).— *Leaves* narrow, wedge-shaped at the base, slightly toothed, sessile ; *stem* round, nearly smooth.—Wet places; frequent.   From 6 to 18 inches high, with very narrow, nearly entire leaves, small flowers, which droop while in bud, and a round stem, which often has 2 downy lines on opposite sides.—Fl. July, August.   Perennial.

* To this group, with undivided stigmas, belong also *E. roseum* (Pale Smooth-leaved Willow-Herb), growing in wet places about London, with stalked, egg-shaped *leaves*, and an imperfectly 4-angled *stem ; E. alsinifolium* (Chickweed-leaved Willow-Herb), a mountainous species, from 6 to 12 inches long, with very thin, egg-shaped, pointed *leaves*, and a few rather large *flowers ;* and *E. alpinum* (Alpine Willow-Herb), also a mountain species, 3 to 4 inches high, which is distinguished from the last by its obtuse *leaves*, and 1 or 2 *flowers*, which droop while in bud.

2. ŒNOTHÉRA (*Evening Primrose*).

1. *Œ. biennis* (Common Evening Primrose).—A tall and stout herbaceous plant, with long, light-green, smooth *leaves*, and large, pale yellow, fragrant *flowers*, which open in the evening, and wither towards the middle of the next day.   It is common in gardens, and

P

in a few places appears to be naturalized.—Fl. July—September. Biennial.

ŒNOTHÉRA BIENNIS (*Common Evening Primrose.*)

3. ISNARDIA.

1. *I. palustris* (Marsh Isnardia).—The only British species.—A small herbaceous plant, 6—8 inches long, with prostrate rooting *stems*, and axillary sessile *flowers*,

which are destitute of *petals.*—Very rare ; in a pool, at Buxtead, Sussex, and on Petersfield Heath, Hampshire. —Fl. July.   Annual.

CIRCÆ LUTETIANA (*Common Enchanter's Nightshade*).

4. CIRCÆA (*Enchanter's Nightshade*).

1. *C. Lutetiána* (Common Enchanter's Nightshade). —*Stem* downy, branched ; *leaves* egg-shaped, tapering to a point, toothed ; *calyx* hairy.—Damp shady places ; common.   A slender herbaceous plant, 1—2 feet high, with creeping roots, spreading foliage, and terminal

clusters of minute white flowers, with pink stamens, which are succeeded by 2-lobed hairy seed-vessels ; often a troublesome weed in damp gardens.—Fl. July, August. Perennial.

2. *C. alpína* (Alpine Enchanter's Nightshade).— *Stem* nearly smooth ; *leaves* heart-shaped, toothed, shining ; *calyx* smooth.—Mountainous woods. Closely resembling the last, but smaller, and less branched. The leaves are remarkable for their delicate texture, and when dried, are nearly transparent.—Fl. July, August. Perennial.

ORD. XXVIII.—HALORÁGEÆ.—The Mare's-tail Tribe.

Calyx adhering to the ovary, and either expanding into 3 or 4 minute lobes, or reduced to a mere rim ; *petals* either minute and placed on the mouth of the calyx, or wanting; *stamens* either equalling the petals in number, twice as many, or, when petals are absent, 1 or 2 ; *ovary* with 1 or more cells ; *stigmas* equal in number to the cells of the ovary ; *capsule* not opening ; *seeds* solitary, pendulous.—An unimportant Order, comprising about 70 species of plants, which are for the most part herbaceous aquatics, with inconspicuous flowers often destitute of petals, and in one genus, *Hippuris* (Mare's-tail), composed of a minute *calyx*, a solitary *stamen*, and a single *seed*. In several species the *stamens* and *pistils* are in separate flowers.

1. HIPPÚRIS (Mare's-tail).—*Calyx* forming a minute, indistinctly 2-lobed rim to the ovary ; *petals* 0 ; *stamen* 1 ; *style* 1 ; *seed* 1, nut-like. (Name in Greek signifying *a mare's tail*.)

2. MYRIOPHYLLUM (Water Milfoil). — *Stamens* and *pistils* in separate flowers, but on the same plant (monœcious) ; *calyx* 4-parted ; *petals* 4 ; *stamens* 8 ; *styles* 4 ; *fruit* of 4 nut-like seeds. (Name from the Greek *myrioi*, 10,000, and *phyllon*, a leaf, from its numerous leaves.)

3. CALLÍTRICHÉ (Water Star-wort).—*Flowers* without calyx or petals, often with 2 bracts at the base; *stamen* 1; *anther* 1-celled ; *styles* 2 ; *ovaries* 2, each 2-lobed ; *fruit* of 4 1-seeded carpels. (Name in Greek signifying *beautiful hair*, from the hair-like, white roots.)

HIPPURIS VULGARIS (*Common Mare's-tail*).

1. HIPPÚRIS (*Mare's-tail*).

1. *H. vulgáris* (Common Mare's-tail).—The only British species, not uncommon in stagnant water. A

singular plant, with erect, jointed *stems*, which are unbranched, except at the base, and taper to a point,
bearing whorls of very narrow *leaves* with hard tips.
The inconspicuous *flowers* are sessile in the axils of the
upper leaves, and are often without stamens.—Not to be
confounded with the genus *Equisétum* (Horse-tail), a
plant allied to the Ferns, which also has a jointed stem
and rigid leaves, but bears its fructification in terminal
spikes, or heads.—Fl. June, July.   Perennial.

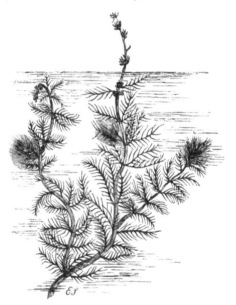

MYRIOPHYLLUM SPICATUM (*Spiked Water Milfoil*).

2. MYRIOPHYLLUM (*Water Milfoil*).

1. *M. spicatum* (Spiked Water Milfoil).—*Flowers* in
whorls, forming a slender leafless spike.—Stagnant

water ; frequent.   An aquatic plant, forming a tangled
mass of slender, much branched stems ; *leaves* 4 in a
whorl, finely divided into numerous hair-like segments, the whole plant being submerged, except the
spikes of inconspicuous greenish flowers, which rise a
few inches above the surface.—Fl. July, August.   Perennial.

     * *M. verticillatum* (Whorled Water Milfoil) differs
from the preceding in having the *flowers* in whorls at
the base of the leaves ; *M. alterniflorum* (Alternate-
flowered Water Milfoil) has the *barren flowers* alternately
arranged in a short leafless spike, with the *fertile flowers*
about 3 together, in the axils of the leaves, at its base.
The last two are rare.

CALLITRICHE VERNA (*Vernal Water Star-wort*).

3. CALLÍTRICHÉ (*Water Star-wort*).

     1. *C. verna* (Vernal Water Star-wort).—*Leaves* in
pairs, united at the base ; *flowers* in the axils of the
leaves ; *carpels* bluntly keeled at the back.—Streams
and stagnant water ; every where.   An aquatic plant
with long slender stems, which send out shining roots
from the joints ; either growing in running water, when
the leaves are usually very narrow, or in stagnant water,
when the upper leaves are broader, and float on the sur-

face, crowded into a starry form, the stamens being the only parts of the plant actually raised above it.—Fl. May—July. Annual.

2. *C. autumnális* (Autumnal Water Star-wort).— *Carpels* winged at the back.—Resembling the last, and growing in similar situations ; but rare. In this species the whole plant is submerged, all the leaves are narrow and abrupt, and of a deeper green.—Fl. June—October. Annual.

* Two other British species of *Callítriché* are described by botanists, which vary in a slight degree from the preceding, but they are not of common occurrence, and are on other accounts scarcely deserving of a separate notice.

Ord. XXIX.—CERÁTOPHYLLEÆ.—The Horn-wort Tribe.

Stamens and *pistils* in separate flowers, but on the same plant (monœcious) ; *calyx* many-parted ; *corolla* none ; *stamens* 12—20, without filaments ; *anthers* 2-pointed ; *ovary* 1-celled ; *style* curved ; *seed-vessel* nut-like, 1-seeded, not opening.—This Order contains only *Cerátophyllum* (Horn-wort), an unimportant family of aquatic plants, very distinct in structure from any other known plants, with rigid whorled *leaves*, which are repeatedly forked, and inconspicuous *flowers*. (Name in Greek signifying *horn-leaved*.)

1. Cerátophyllum (*Horn-wort*).

1. *C. demersum* (Common Horn-wort).—*Fruit* armed with 2 thorns near the base, and terminated by the curved *style*.—Slow streams and ditches ; frequent. An aquatic plant, growing entirely under water ; with long, slender *stems ;* whorled, bristle-like *leaves*, which are 2—4 times forked, and often inflated and jointed ; the *flowers* also are whorled, and grow in the axils of the leaves.—Fl. July. Perennial.

* *P. submersum* scarcely differs from the preceding, except in having fruit without thorns.

CERATOPHYLLUM DEMERSUM (*Common Horn-wort*).

Ord. XXX.—LYTHRÁRIÆ.—Loosestrife Tribe.

Calyx tubular, many-parted, often with intermediate teeth ; *petals* inserted between the outer divisions of the calyx, soon falling off ; *stamens* springing from the tube of the calyx, within the petals, and either equalling them in number, or twice, thrice, or four times as many ; *ovary* 2—4-celled ; *style* single ; *capsule* many-seeded, covered by the calyx, but not united to it.—Herbaceous plants, mostly with opposite leaves, and 4-cornered stems, inhabiting Europe, America, and India. Many of the plants of this tribe possess astringent properties, and some are used for dyeing. The common Purple Loosestrife is found in Australia. *Lawsonia inermis* is the plant from which the Henna of Egypt is obtained. It is used by the women of that country to stain their nails of an orange colour, and is also employed for dyeing Morocco leather reddish-yellow.

1. LYTHRUM (Purple Loosestrife).—*Calyx* cylindrical, with 12 divisions, alternately smaller; *petals* 6; *stamens* 6 or 12; *style* thread-like. (Name from the Greek *lythron*, blood, from the colour of the flowers.)

2. PEPLIS (Water Purslane).—*Calyx* bell-shaped, with 12 divisions, alternately smaller; *petals* 6, minute, soon falling off; *stamens* 6; *style* very short. (Name of Greek origin, and anciently applied to another plant.)

PEPLIS PÓRTULA (*Water Purslane*).

LYTHRUM SALICARIA (*Purple Loosestrife, or Willowstrife*).

1. LYTHRUM (*Purple Loosestrife*).

1. *L. salicaria* (Purple Loosestrife, or Willow-strife).
—*Leaves* opposite, long, and narrow, heart-shaped at

the base ; *flowers* whorled, in leafy spikes ; *stamens* 12.
—Watery places ; abundant. An exceedingly handsome plant, 2—4 feet high, generally growing on river
banks, among sedges and rushes, and sending up tall,
tapering spikes of purple flowers, which seen from a distance might be mistaken for Foxgloves.—Fl. July,
August. Perennial.

* *L. hyssopifolia* (Hyssop-leaved Purple Loosestrife)
is a much smaller plant, 4—6 inches high, with alternate *leaves*, and solitary purple *flowers*, with 6 *stamens*.
It grows in moist places, but is far from common.

2. PEPLIS (*Water Purslane*).

1. *P. Pórtula* (Water Purslane).—A humble, creeping, aquatic plant, with opposite smooth *leaves*, and
inconspicuous axillary *flowers*. The *stems* are usually
more or less tinged with red ; and when the plant grows
in places from which the water has been dried up, the
leaves acquire the same hue. — Fl. July, August.
Annual.

ORD. XXXI.—TAMARISCINEÆ.—THE TAMARISK TRIBE.

Calyx 4—5 parted, overlapping when in bud, remaining after the petals have withered ; *petals* 4—5, from
the base of the calyx ; *stamens* either equal to the petals
in number, or twice as many, distinct, or united by their
filaments ; *ovary* not combined with the calyx ; *styles* 3 ;
capsule 3-valved, 1-celled, containing many seeds, which
are tufted with down at the extremity.—Mostly shrubs,
with rod-like branches and minute leaves, which resemble scales. They are found only in the eastern half
of the northern hemisphere, and are most numerous on
the shores of the Mediterranean ; but though preferring
the sea-side, they are not unfrequently found on the
banks of rivers, and occur, also, in the desert, especially
where the soil is impregnated with salt, as in the neigh-

bourhood of Mount Sinai, where a species of Tamarisk very like the common one, produces a sugary substance, called by the Arabs *Manna*. The bark is astrin-

TAMARIX GALLICA (*Common Tamarisk*).

gent, and several species are remarkable for the large quantity of sulphate of soda contained in their ashes, and for the galls which they bear on their branches.

These are highly astringent, and are used both in
medicine and in dyeing. For a further account of the
Manna produced by the Tamarisk of Mount Sinai, see
" Forest Trees of Britain."

1. TÁMARIX (Tamarisk).—*Calyx* 5-parted ; *petals* 5 ;
stamens 5 or 10 ; *stigmas* feathery. (Named from the
Tamarisci, a people who inhabited the banks of the
Tamaris, now Tambra, in Spain, where the Tamarisk
abounds.)

TÁMARIX (*Tamarisk*).

1. *T. Gállica* (Common Tamarisk). — A handsome
shrub or small tree, with long flexible *branches*, and
minute scale-like *leaves*, which are close pressed to the
twigs, and give the tree a light feathery appearance.
The *flowers*, which are rose-coloured, grow in clusters of
spikes. Some botanists have named this species *Támarix
Ánglica* (English Tamarisk); but even if it be a distinct
species, it can scarcely be considered a native of Britain,
occurring only in a few places on the south-western coast,
and everywhere appearing to have been planted.—Fl.
July. Shrub.

ORD. XXXII.—CUCURBITACEÆ.—THE GOURD
TRIBE.

Stamens and *pistils* in separate flowers, either on the
same plant (Monœcious) or on different plants (Diœcious) ; *calyx* 5-toothed, united with the corolla ; *corolla*
often scarcely to be distinguished from the calyx ; *stamens* 5, more or less united ; *anthers* twisted ; *ovary*
imperfectly 3-celled ; *style* short ; *stigmas* short, thick,
lobed, velvety ; *fruit* more or less juicy ; *seeds* flat, wrapped in a skin.—A large and important Order, containing herbaceous plants with juicy stems, and climbing
by means of tendrils, which arise from the base of the
leaf-stalks. The leaves are usually lobed and rough ;

the flowers often large, white, red, or yellow ; the fruit
juicy, or fleshy. They inhabit principally the hot
regions of the globe, but a few are found in temperate
climates ; and a great number are cultivated in Europe,
either for ornament or use. Their properties are in
many instances exceedingly violent, of which the common drug *Colocynth* affords an example ; the Bottle
Gourd is another, it being recorded that some sailors
were poisoned by drinking beer that had been standing
in a flask made of one of these gourds. The poisonous
plant mentioned in 2 Kings iv. 39, 40, is supposed to
be a plant of this tribe, the *Wild*, or *Ass Cucumber*,
which bears an oval fruit of a very bitter taste, and grows
in sandy and desert places. As this cucumber has very
much the same appearance as that which is cultivated
in gardens, but only is somewhat smaller, and as even
its leaves and tendrils are similar, it might easily happen that the man sent out by the disciples of the pro-
phets took wild Cucumbers for a harmless fruit, and
prepared a meal of them. But the bitter taste of the
boiled Cucumber made those who tasted it fear that it
was poisonous, the opinion being general with the
Hebrews that bitter taste indicated the presence of poison, (see Rev. viii. 10, 11.) The only plant belonging
to this tribe which is a native of Britain, *Bryonia dioïca*
(White Bryony), partakes of the properties of Colocynth,
and the root is said to be a valuable medicine. The
Spirting Cucumber, so called from the force with which
it expels the poisonous juice when ripe, is a very dangerous drug ; a few grains of Elaterium, a prepared form
of this juice, having been known to bring on symptoms
of poisoning. A case even is recorded, when a person
was taken dangerously ill from having merely carried a
specimen in his hat. Many species, however, produce
edible fruit ; for instance, the numerous varieties of
Melon and Cucumber, the Water Melon, so highly
esteemed for the cool refreshing juice of its ripe fruit,
and one of our finest table vegetables, the Vegetable

Marrow. It is said that the tender shoots of the White Bryony may be used with safety after having been boiled, and that they resemble Asparagus in flavour; but it is highly probable that the shoots of Black Bryony (*Tamus*

BRYONIA DIOÍCA (*White Bryony*).

communis), a plant belonging to a different Order, may have been used instead; in either case the experiment is a dangerous one.

1. BRYONIA (Bryony).—*Stamens* 5, in 3 sets; *style* 3-cleft; *fruit* a globose berry. (Name from the Greek

bryo, to bud, the rapid growth of the Gourd tribe being proverbial.)

1. Bryonia (*Bryony*).

1. *B. dioica* (White Bryony).—The only British species; frequent in England, except in the extreme western counties. An elegant climbing plant, with large, light green, rough *leaves*, having undivided tendrils at the base, and bunches of whitish *flowers*, with green veins. The *fertile flowers* may be distinguished at once from the *barren*, by the presence of an *ovary* below the calyx. These are succeeded by globular scarlet *berries*, which hang about the bushes after the stems and leaves have withered. The berries of Black Bryony (*Tamus communis*) are larger, and elliptical in shape ; both should be avoided as injurious, if not poisonous.—Fl. May— August. Perennial.

Ord. XXXIII.—PORTULÁCEÆ.—The Purslane Tribe.

Calyx of 2 sepals, united at the base ; *petals* usually 5, from the base of the calyx ; *stamens* 3 or more, inserted with the petals; *ovary* 1-celled ; *style* 1 or 0 ; *stigmas* several ; *capsule* 1-celled, opening transversely, or by 3 valves ; *seeds* usually more than 1.—Herbs or shrubs, with juicy stems and leaves, and irregular flowers, which open only during sunshine. The most remarkable plant in this Order is the common Purslane, which has been used from all antiquity as a pot-herb. Many species have large, showy flowers, but the only one which occurs in Britain is a humble plant, with very inconspicuous flowers.

1. Montia (Water Blinks).—*Calyx* of 2 sepals ; *corolla* of 5 petals, 3 smaller than the others, and all united at the base ; *tube of the corolla* split to the base ; *capsule* 3-valved, 3-seeded.

Q

MONTIA FONTANA (*Water Blinks*).

1. MONTIA (*Water Blinks*).

1. *M. fontána* (Water Blinks).—The only species ; abundant in wet places.   An unpretending little plant, well marked by its 2-leaved *calyx*, irregular *corolla*, the tube of which is split in front, and by its *capsules* containing 3 dotted *seeds.*—Fl. June—August.   Annual.

ORD. XXXIV.—PARONYCHIEÆ.—THE KNOT-GRASS TRIBE.

Sepals usually 5 ; *petals* 5, minute, inserted between the lobes of the calyx, sometimes wanting ; *stamens* varying in number, opposite the petals, if equalling them in number ; *ovary* not combined with the calyx ; *pistils* 2—5 ; *fruit* 1-celled, opening with 3 valves, or not at all.—Small, branching, herbaceous, or somewhat shrubby plants, with sessile undivided leaves and minute flowers, principally confined to the south of Europe, and north of Africa, where they grow in the most barren places, covering with a thick vegetation soil which is incapable of bearing anything else.   A few only are found so far north as Great Britain, and nearly all of these are confined to the southern shores.

1. CORRIGÍOLA (Strapwort).—*Sepals* 5 ; *petals* 5, as long as the calyx ; *stamens* 5 ; *stigmas* 3, sessile ; *fruit* 1-seeded, enclosed in the calyx. (Name from *corrigia*, a strap, from the shape of the leaves.)

CORRIGIOLA LITTORALIS (*Sand Strapwort*).

2. HERNIARIA (Rupture-wort).—*Sepals* 5 ; *petals* 5, resembling barren filaments ; *stamens* 5, inserted on a fleshy ring ; *stigmas* 2, nearly sessile ; *fruit* 1-seeded, enclosed in the calyx. (Name from the disease for which the plant was formerly supposed to be a remedy.)

3. ILLÉCEBRUM (Knot-grass).—*Sepals* 5, coloured, thickened, ending in an awl-shaped point ; *petals* 0, or 5 ; *stigmas* 2 ; *fruit* 1-seeded, enclosed in the calyx. (Name from the Latin *illécebra*, an attraction.)

4. POLYCARPON (All-seed). — *Sepals* 5 ; *petals* 5, notched ; *stamens* 3—5 ; *stigmas* 3, nearly sessile ; *fruit* 1-celled, 3-valved, many-seeded. (Name from the Greek *polys*, many, and *carpos*, fruit.)

5. SCLERANTHUS (Knawel).—*Calyx* 5-cleft, contracted at the mouth of the tube ; *petals* 0 ; *stamens* 10, rarely

5 ; *styles* 2 ; *fruit* 1-seeded, covered by the hardened calyx. (Name from the Greek *scléros*, hard, and *anthos*, a flower, from the hardness of the calyx.)

1. CORRIGÍOLA (*Strapwort*).

1. *C. littorális* (Sand Strapwort).—A small but pretty plant, with slender spreading *stems*, which lie quite prostrate, very narrow strap-shaped *leaves*, of a glaucous hue, and tufts of small white *flowers*.—It grows in two or three places on the sea-shore of Devon, and is very abundant on the banks of the Loe Pool, near Helston, Cornwall.—Fl. August—October.   Annual.

2. HERNIARIA (*Rupture-wort*).

1. *H. glabra* (Smooth Rupture-wort).—A small prostrate plant, with much of the habit of Wild Thyme ; abundant in the neighbourhood of the Lizard Point, Cornwall, but very rare elsewhere. Though called *smooth*, the leaves are always more or less hairy, and sometimes strongly fringed at the edges.   The flowers are green, and grow in sessile tufts in the axils of the leaves, or, not unfrequently, crowded into leafy spikes.   Some botanists consider the more hairy variety a distinct species, but without sufficient reason ; as the plant assumes various forms, according to soil, and its being more or less exposed.—Fl. July—September. Perennial.

3. ILLÉCEBRUM (*Knot-grass*).

1. *I. verticillatum* (Whorled Knot-grass).—In boggy ground, and standing water, among other aquatic plants ; in Cornwall, not uncommon ; much rarer in Devonshire. A pretty plant, with slender tangled *stems*, of a red tint, glaucous, sessile *leaves*, and axillary whorls of white *flowers*, which are remarkable for their thickened calyx-leaves terminating in a soft point.—Fl. July—September. Perennial.

4. POLYCARPON (*All-seed*).

1. *P. tetraphyllum* (Four-leaved All-seed).—On the southern coast, but far from common. A small plant, with prostrate, branched *stems*, and many minute greenish-white *flowers*, with 3 stamens. The lower *leaves* grow in fours, the upper in pairs.—Fl. May—August. Annual.

SCLERANTHUS ANNUA (*Annual Knawel*).

5. SCLERANTHUS (*Knawel*).

1. *S. annua* (Annual Knawel).—*Calyx*, when in fruit, spreading, acute, with a narrow whitish margin ; root annual.—Corn-fields, especially on a gravelly soil ; common. A small plant, 2—4 inches high, with numerous tangled stems, awl-shaped leaves, and green flowers,

which grow either in the forks of the stems, or in terminal tufts.—Fl. July—November. Annual.

* *S. perennis* (Perennial Knawel) has prostrate *stems*, blunt *calyx-leaves*, with a broad margin, and perennial *roots*. It grows in dry sandy fields, in Norfolk and Suffolk.

Ord. XXXV.—CRASSULACEÆ.—The Stonecrop Tribe.

Sepals 3—20, more or less united at the base ; *petals* equal to the sepals in number, inserted in the bottom of the calyx ; *stamens* the same, or twice as many, in which latter case, those opposite the petals are shorter than the others ; *ovaries* as many as petals, 1-celled, tapering into stigmas, often with a gland at the base of each ; *fruit* consisting of several erect seed-vessels, which open lengthwise ; *seeds* in a double row.—Herbs or shrubs, remarkable for their thick, fleshy leaves, and starlike flowers, inhabiting most parts of the world, especially the south of Africa, and growing in the driest situations, where not a blade of grass nor a particle of moss can live ; on naked rocks, old walls, or sandy, hot plains,—alternately exposed to the heaviest dews of night, and the fiercest rays of the noon-day sun,—requiring no nourishment, save what they derive from the atmosphere through myriads of mouths invisible to the naked eye, but covering all their surface. A common British species, *Sedum Teléphium,* "Orpine living long," will grow for months, if suspended by a string from the ceiling of a room, without being once supplied with water. An African species, *Bryophyllum calycinum,* will not only grow if similarly treated, but if its leaves be gathered and laid on the ground, they will send out from the notches on their margin, young shoots, in all respects resembling the parent plant. The properties of the tribe are in general acrid ; some few contain malic acid, and one or two are sometimes used in medicine, for their astringent properties.

1. TILLÆA.—*Sepals, petals, stamens,* and *carpels,* 3 or 4, the latter 2-seeded. (Named after *Michael Angelo Tilli,* an Italian botanist.)

2. COTYLÉDON (Pennywort).—*Sepals* 5 ; *corolla* tubular, 5-cleft ; *carpels* 5, with a scale at the base of each. (Name from the Greek *cótylé,* a cup, from the shape of the leaves.)

3. SEMPERVÍVUM (House-leek).—*Sepals, petals,* and *carpels,* 6—10 ; *stamens* twice as many. (Name from the Latin *semper,* always, and *vivo,* to live.)

4. SEDUM (Stonecrop).—*Sepals* and *petals* 5 ; *stamens* 10, spreading ; *carpels* 5. (Name from the Latin *sedeo,* to sit ; from the humble growth of the plants.)

5. RHODÍOLA (Rose-root).—*Stamens* and *pistils* on different plants ; *sepals* and *petals* 4 ; *stamens* 8 ; *carpels* 4. (Name from the Greek *rhodon,* a rose, from the scent of the roots.)

1. TILLÆA.

1. *T. muscósa* (Mossy Tillæa).—A minute plant, with small, opposite, blunt *leaves,* and greenish-white *flowers* tipped with red ; more frequent as a troublesome weed in gravel walks, than elsewhere. It has somewhat of the habit of a *Sagína,* from which, however, it is very distinct.—Fl. May, June. Annual.

2. COTYLÉDON (*Pennywort*).

1. *C. umbilícus* (Wall Pennywort).—A remarkably succulent plant, with circular, notched *leaves,* which are more or less sunk above, and are supported on their stalks by their centres, or *peltate.* The *flowers* are pendulous, and grow in spike-like clusters, 6—18 inches high, of a greenish yellow colour. The leaves are well

known to children by the name of "penny pies."—
Fl. June—August.   Perennial.

COTYLEDON UMBILÍCUS (*Wall Pennywort*).

3. SEMPERVÍVUM (*House-leek*).

1. *S. tectórum* (Common House-leek).   A common,
but scarcely indigenous plant, growing on the roofs of
cottages.   The *leaves* are thick and juicy, fringed at the
edges, and grow in compact rose-like tufts.   Each of the

purple *flowers* contains 12 perfect, and 12 imperfect *stamens;* the latter, which are arranged alternately with the petals, frequently bearing *anthers* containing embryo *seeds,* like those found in the carpels, but of course never attaining maturity. The leaves contain malic acid, the same acid which is found in the apple.—Fl. July. Perennial.

SEMPERVIVUM TECTORUM (*Common House-leek*).

4. SEDUM (*Stonecrop*).

* *Leaves flat ; flowers purple.*

1. *S. Telephium* (Orpine, or Live-long). — *Leaves* oblong egg-shaped, serrated ; *stems* erect.—The largest British species, and well distinguished, not only by its terminal corymbs of purple flowers, but by its large, broad leaves. It grows about 2 feet high, in bushy places, and is remarkable for the length of time that it will continue fresh after being gathered ; whence it derives its name, *Live-long.* — Fl. July, August. Perennial.

** *Leaves scarcely, if at all, flattened ; flowers white.*

2. *S. Anglicum* (English Stonecrop).—*Leaves* egg-shaped, fleshy, spurred at the base beneath, sessile ; *cymes* 2-cleft ; *petals* very sharp.—Rocky and sandy places, especially near the sea.—A small plant, 3—4 inches high, with stems which at first are prostrate and rooting, afterwards erect ; the leaves are mostly alternate, often tinged with red, small, and very thick ; the flowers conspicuous from their star-like form, their white petals spotted with red, and bright purple anthers.— Fl. June, July. Annual.

3. *S. album* (White Stonecrop). — *Leaves* oblong, cylindrical, blunt, spreading ; *cyme* much branched, drooping when in bud.—Rocks and walls; not common. Rather taller and more slender than the last, and easily distinguished by its much longer cylindrical leaves, and more numerous white, not spotted, flowers. This species is not uncommon on garden walls, and the roofs of outhouses.—Fl. July, August. Perennial.

* To this group belong *S. villósum* (Hairy Stonecrop), a small species, with hairy, viscid *stems* and *leaves,* and pinkish white *flowers;* frequent in Scotland and the north of England : and *S. dasyphyllum* (Thickleaved Stonecrop), also a small species, but very rare ;

distinguished from the preceding by its fleshy, almost globular *leaves*, viscid *flower-stalks*, and blunt *petals*.

*** *Leaves scarcely, if at all, flattened; flowers yellow.*

4. *S. acre* (Biting Stonecrop).—*Leaves* egg-shaped, fleshy, spurred at the base, sessile; *cymes* 3-cleft.— Walls, rocks, and sandy ground; frequent. Very like *S. Ánglicum* in habit, but with yellow flowers, and growing in similar situations; it may, however, be distinguished, when not in flower, by its thicker and more crowded leaves, which are very acrid, and have gained for the plant the name of *Wall-pepper.*—Fl. June, July. Perennial.

SEDUM ACRE (*Biting Stonecrop*).

5. *S. reflexum* (Crooked Yellow Stonecrop).—*Leaves* awl-shaped, spurred at the base, nearly cylindrical; the

lowermost curved back.—Walls and dry banks; not uncommon. Easily distinguished from any of the preceding by its slender, but tough stems, 6—10 inches high, clothed with spreading, or reflexed leaves, which are cylindrical and pointed.—Fl. July, August. Perennial.

RHODIOLA ROSEA (*Rose-root*).

Other British species, belonging to this group, are *S. sexangulare* (Tasteless Yellow Stonecrop), distinguished from *S. acre* by its *leaves*, which are 6 in a whorl, growing in Greenwich Park, the Isle of Sheppey, and a few other places: *S. rupestre* (St. Vincent's Rock Stonecrop), a species allied to *S. reflexum*, with slightly flattened *leaves*, which grow 5 in a whorl, found on St. Vincent's Rocks, and other limestone cliffs; rare: and *S. Forsterianum* (Welsh Stonecrop), another species allied to *S. reflexum*, with *leaves* flattened at the base, and compact *cymes* of flowers, which grows on rocks in Wales and Shropshire.

5. RHODÍOLA (*Rose-root*).

1. *R. rosea* (Rose-root).—The only species. A succulent, broad-leaved plant, with the habit of *Sedum Teléphium* (Orpine), but stouter.—Abundant on mountains in Scotland, Ireland, and the north of England, and found also on sea-cliffs. The *flowers* are greenish yellow, and grow in compact terminal *cymes*, on simple *stems*, 6—10 inches high; *roots* thick and knotted, having the perfume of rose-water.—Fl. June. Perennial.

ORD. XXXVI.—GROSSULARIEÆ.—THE GOOSEBERRY AND CURRANT TRIBE.

Calyx growing from the summit of the ovary, 4—5 cleft; *petals* 4—5, small, inserted at the mouth of the calyx-tube, and alternating with the stamens; *ovary* 1-celled, with the *ovules* (young seeds) arranged in two opposite rows; *style* 2—4-cleft; *berry* crowned with the withered flower, pulpy, containing many stalked seeds. —Shrubs with or without thorns, and having alternate lobed leaves, which are plaited when in bud. The flowers grow in clusters in the axils of the leaves, each flower with a bract at its base, and are succeeded by pulpy berries, which in several species are highly prized for their agreeable flavour. In other species the taste is mawkish, or extremely acid. The plants of this tribe grow only in the temperate parts of the world especially in North America, and on the mountains of Northern India. In Africa they are unknown.

1. RIBES (Currant and Gooseberry).—*Calyx* 5-cleft ; *petals* 5, inserted at the mouth of the calyx-tube ; *stamens* 5 ; *berry* many-seeded, crowned by the withered flower. (Name anciently given to a species of Rhubarb.)

1. RIBES (*Currant and Gooseberry*).

* *Flowers* 1—3 *together ; branches thorny.*

1. *R. grossularia* (Gooseberry).—The common goose-berry of gardens. Frequently met with in hedges and thickets, but not considered to be a native plant. It is well distinguished by its sharp *thorns*, which grow either singly, or 2—3 together, below the leaf-buds.— Fl. April, May. Shrub.

** *Flowers in clusters ; branches without thorns.*

2. *R. rubrum* (Red Currant).—*Clusters* drooping ; *bracts* very small ; *calyx* smooth ; *leaves* bluntly 5-lobed. —The Red and White Currant of gardens ; not un-common in hedges near houses ; and in Scotland and the north of England supposed to be wild.—Fl. April, May. Shrub.

3. *R. nigrum* (Black Currant).—Clusters loose, droop-ing, with a single stalked flower at the base of each ; *calyx* downy ; *leaves* sharply 3—5-lobed, dotted with glands beneath.—The Black Currant of gardens ; occa-sionally wild in damp woods. Easily distinguished, at all seasons, by the strong perfume of its buds and leaves.—Fl. April, May. Shrub.

* *R. alpinum* (Tasteless Mountain Currant) has the *stamens* and *pistils* on separate plants ; the *flowers* also grow in erect clusters, with very long *bracts* at the base of each. It grows in the mountainous parts of England and Scotland.

RIBES NIGRUM (*Black Currant*).

Ord. XXXVII.—SAXIFRAGEÆ.—The Saxifrage Tribe.

Sepals 5, rarely 4, more or less united at the base ; *petals* equalling the *sepals* in number, inserted between the sepals, rarely 0 ; *stamens* 5—10 ; *ovary* of two united carpels ; *styles* 2, usually spreading in opposite directions ; *capsule* 2-celled, opening on the inside ; *seeds* numerous. This order is principally composed of herbaceous, mountainous plants, with tufted foliage, and glandular stems. They abound in temperate and cold climates, but are not found in tropical countries. The genus *Saxifraga* (Saxifrage) is an extensive one, and contributes greatly to the beauty of the vegetation high up in the mountains ; but some species grow on old walls, by the sides of rivulets, and in moist meadows. *Chrysosplenium* (Golden Saxifrage) has no petals. Few of the plants belonging to this tribe are applied to any use. Most of them have slight astringent properties, and some few are bitter and tonic.

1. SAXIFRAGA (Saxifrage).—*Calyx* in 5 divisions ; *petals* 5 ; *stamens* 10 ; *styles* 2 ; *capsule* 2-celled, 2-beaked, opening between the beaks ; *seeds* numerous. (Name in Latin signifying *rock-breaker*, many of the species growing in crevices of rocks.)

2. CHRYSOSPLENIUM (Golden Saxifrage).—*Calyx* in 4 divisions ; *petals* 0 ; *stamens* 8, rarely 10 ; *styles* 2 ; *capsule* 2-beaked. (Name from the Greek, *chrysos*, gold, and *splen*, the spleen, from some imaginary virtues of the plant.)

1. SAXÍFRAGA (*Saxifrage*).

* *Calyx reflexed, inferior ; flowers whitish, panicled.*

1. *S. stelláris* (Starry Saxifrage) —*Leaves* oblong, wedge-shaped, toothed, scarcely stalked ; panicle of few flowers.—Wet rocks, and sides of rivulets, in Scotland, Ireland, and the north of England. A mountain plant, 3—5 inches high, with coarsely toothed leaves, and rather large white petals, each with two yellow spots near the base.—Fl. June, July. Perennial.

2. *S. umbrósa* (London Pride, or St. Patrick's Cabbage).—*Leaves* roundish, egg-shaped, with white notches, tapering at the base into a flat stalk.—In the south and west of Ireland, plentiful ; naturalized in many parts of England, and very common in gardens. A well-known plant, with rose-like tufts of fleshy leaves, and panicles of small white flowers, dotted with pink. Though growing naturally on mountains, there is scarcely any situation where it will not make itself at home, even in the smoky gardens of London. Hence it varies considerably in form, and has been subdivided by some botanists into several species.—Fl. June. Perennial.

* To this group belong *S. Géum* (Kidney-shaped
Saxifrage), distinguished by its kidney-shaped *leaves;*
common on mountains in the south of Ireland : and
S. hirsúta, a species intermediate between *S. Géum* and
S. umbrósa, also an Irish plant, but very rare.

** *Calyx spreading; leaves not divided.*

3. *S. nivális* (Clustered Alpine Saxifrage).—*Leaves*
all from the root, inversely egg-shaped, sharply crenate;
calyx half inferior; *flowers* in a crowded head.—Mountains in Wales and Scotland. An alpine plant, 3—6
inches high, with rather large, white flowers, which
grow in a compact head.

4. *S. aizoídes* (Yellow Mountain Saxifrage).—*Leaves*
very narrow, fleshy, fringed; *flowers* in a leafy panicle.
—Wet places in the mountains; abundant. A handsome species, about 6 inches high, with large, bright
yellow flowers, spotted with scarlet.—Fl. June—September. Perennial.

* *S. Hírculus* (Yellow Marsh Saxifrage) differs from
the last in having its *flowers* solitary, or nearly so; it is
of very rare occurrence : *S. oppositifolia* (Purple Mountain Saxifrage) has egg-shaped *leaves*, and large, solitary, purple *flowers;* it grows on the mountains of
Wales, Yorkshire, and Scotland.

*** *Calyx spreading; leaves divided.*

5. *S. granuláta* (White Meadow Saxifrage).—*Root-
leaves* kidney-shaped, with rounded lobes, stalked;
stem-leaves nearly sessile, sharply lobed; *flowers* panicled; *roots* granulated.—Gravelly meadows; not uncommon. A pretty plant, with slender, leafy stems,
10—12 inches high, and rather large, pure white
flowers. The roots are remarkable for producing numerous, downy, bulb-like tubes. A double variety is
frequent in gardens.—Fl. May, June. Perennial.

R

SAXIFRAGA GRANULATA (*White Meadow Saxifrage*).

6. *S. tridactylites* (Rue-leaved Saxifrage). Whole plant viscid with glandular hairs; *leaves* wedge-shaped, 3—5-cleft; *stem* much branched; *flowers* terminal, each on a separate stalk.—On the tops of walls, and roofs of cottages; common. A small species, rarely more than 3 inches high, with very hairy and viscid stems, and small white flowers. The whole plant has usually a red tinge.—Fl. May, June. Annual.

7. *S. hypnoides* (Mossy Saxifrage).—*Root-leaves* 3—5-cleft; those on the creeping-shoots 3-cleft, or entire;

lobes of the leaves all very narrow, acute, bristle-pointed, and fringed.—Mountainous places; common. Distinguished by its dense tufts of finely divided leaves, and loose panicles of rather large, white flowers. Very frequent in gardens.—Fl. May—July. Perennial.

* Closely allied to the preceding is *S. cæspitosa* (Tufted Alpine Saxifrage), distinguished by the obtuse lobes of its *leaves*, and by other minute characters : *S. cernua* (Drooping Bulbous Saxifrage) is a very rare species, which grows only on dry rocks on the summit of some of the Highland Mountains; it is remarkable for bearing *bulbs* in the axils of the *leaves*, and a solitary terminal white *flower: S. rivularis* (Alpine Brook Saxifrage), another very rare species, grows on moist rocks among the same mountains : besides which, other species and varieties have been described, but an enumeration of them is not thought necessary here.

CHRYSOSPLENIUM OPPOSITIFOLIUM *(Common Golden Saxifrage).*

2. CHRYSOSPLENIUM *(Golden Saxifrage).*

1. *C. oppositifolium* (Common Golden Saxifrage).— *Leaves* opposite, roundish heart-shaped.—Sides of shady

rivulets, and wet woods; common. A small aquatic
plant, about 6 inches high, with abundance of bright
green, tender foliage, and terminal, flat clusters of yellowish green flowers.—Fl. April, May.   Perennial.
    2. *S. alternifolium* (Alternate-leaved Golden Saxi-
frage).—*Leaves* alternate, lower ones kidney-shaped, on
long stalks. —Very like the last, and growing in similar
situations, but rare.   The flowers in this species are of
a deeper yellow ; in both, the number of stamens is
usually 8.—Fl. April, May.   Perennial.

Ord. XXXVIII.—UMBELLIFERÆ.—The Umbelliferous Tribe.

Calyx superior, 5-toothed, often reduced to a mere
margin ; *petals* 5, usually ending in a point, which is
bent inwards ; *stamens* 5, alternate with the petals,
curved inwards when in bud ; *ovary* inferior, 2-celled,
crowned by a fleshy disk, which bears the petals and
stamens ; *styles* 2 ; *stigmas* small ; *fruit* composed of
2 carpels, which adhere by their faces to a central
stalk, from which as they ripen they separate below,
and finally are attached by the upper extremity only ;
each *carpel* is marked by 5 vertical ridges, with 4 intermediate ones ; these ridges are separated by channels,
in which are often found, imbedded in the substance of
the fruit, narrow cells (called *vittæ*), containing a coloured oily matter ; *seeds* 1 in each carpel, attached by
their upper extremity, and containing a large horny
albúmen; the flowers are usually small, and situated on
the extremities of little stalks, which are united at the
base, and form an *umbel*. When several of these smaller
umbels proceed in like manner from a common stalk,
the umbel is said to be *compound;* the larger being
called a *general umbel*, the smaller *partial*.   The small
leaves which commonly accompany the flowers of this
tribe are called *general*, or *partial bracts*, according to

their position; each collection of bracts is sometimes termed an *involucre.*—All the British plants belonging to this order are herbaceous, with tubular, or solid, jointed stems. With two exceptions, *Eryngium* and *Hydrocótylé*, they have compound umbels. By far the larger number have also divided leaves, more or less sheathing at the base, and white flowers. Though it is easy to decide at a glance to what *order* they are to be assigned, no such facility exists in distinguishing the *families* of the Umbelliferæ. Indeed, were it not for the large number of species (1,500), which are known to exist, it is probable that they would have been brought together by botanists, so as to form but a few genera, whereas they have been divided into as many as 267; and, as all these agree in the more important parts of fructification, the distinctions of the genera are necessarily founded on differences so minute, that, in the case of other plants, they would perhaps be considered sufficient to do no more than distinguish species. To the young botanist, therefore, the study of the Umbelliferæ is unusually difficult; all the more important distinctions being founded on the ripe fruit, namely—the number, position, and shape of the *ridges*—the presence or absence of *vittæ*—and the form of the *albúmen.* As it would be absurd, in a work professing to be a popular description of British Wild Flowers, to attend solely, or even in any great degree, to these characters, it has been thought desirable to limit the number of species described to those which are of most common occurrence, and to notice any peculiarity in growth, which, though not strictly admissible into a systematic description, may assist the student in discerning the names of the plants he may meet with. For a fuller list, and for more accurate information, he is referred to works of a professedly scientific character, such as *Hooker's British Flora,* and *Babington's Manual of British Botany.*

Among the large number of species of which this tribe is constituted, one would naturally expect to find plants varying greatly in their properties. And such is the case to a certain extent; the roots, leaves, and seeds are variously employed; some as food and condiments, others as medicine, while others are highly poisonous. Yet, when considered with reference to their properties, they may be conveniently arranged into 5 groups; all the members of each group being remarkably similar. The first comprises plants which abound in an acrid, watery juice, which is more or less narcotic in its effects on the animal frame, and which, therefore, when properly administered in minute doses, is a valuable medicine. Among these, the most important is *Conium* (Hemlock); every part of this plant, especially the fresh leaves and green fruit, contains a volatile, oily alkali, called *Conia*, which is so poisonous that a few drops soon prove fatal to a small animal. It acts on the nervous system, and is a valuable medicine in cancerous and nervous diseases. Several other British species are poisonous, especially *Œnanthé, Cicúta*, and *Æthúsa*, described below. The second group comprises those which abound in a resinous gum, of a fetid odour, which is supposed to be owing to the presence of sulphur in combination with the peculiar essential oil. Among these, the first place is held by *Asafœtida*, the hardened milky juice of various species of *Férula*, inhabiting Persia and the neighbouring countries. This drug was held in high repute among the ancients for its medical virtues; it was supposed to be an antidote to poison, to restore sight to the blind, and youth to the aged; and was besides considered a certain specific against various diseases. Gum Gálbanum is the produce of other umbelliferous plants, natives of the East. The third group comprises plants the seeds of which abound in a wholesome aromatic oil. The principal of these are well known, under the names of *Caraway, Coriander, Dill, Anise*, and *Cumin*. The fourth group comprises plants which contain some of the above properties in a very slight degree, or so modified as to form wholesome esculent vegetables. Among these, *Carrots* and *Parsneps* occupy the first place ; *Celery* and *Alexanders*, in their wild state, are too acrid to be used as food, but, when blanched by artificial means, become mild and agreeable; *Parsley, Fennel*, and *Chervil*, the last now nearly out of use, are well-known potherbs ; *Samphire* affords the best of pickles ; the root of *Eryngo* is sweet, aromatic, and tonic, and is commonly sold in a candied state ; the root of *Angelica* (*Angélica Archangélica*) is fragrant and sweet when first tasted, but leaves a glowing heat in the mouth, and is commended by the Laplanders both as food and medicine ; the candied stems form a favourite sweetmeat. Several species produce underground tubers, which, under the name of *pig-nuts*, or *earth-nuts*, are eaten by children and pigs ; and others, common in the East, afford valuable pasturage for cattle. Of all the British umbelliferous plants, the most dangerous are the Water Dropworts (*Œnanthé*), the large tuberous roots of which, resembling Dahlia roots, are often exposed by the action of running water, near which they grow, and are thus easily got at by children and cattle.

The following table contains a description of all the common British species: a list of the rarer ones, and of introduced species, will be found at the end of the Order.

* Umbels simple, or irregular.

1. HYDROCÓTYLÉ (White-rot).—*Flowers* in simple umbels ; *fruit* of 2 flattened, roundish lobes, united by the narrow edge; *leaves* round, peltate. (Name from the Greek, *hydor*, water, and *cótylé*, a platter, from the shape of the leaves, and place of growth.)

2. SANÍCULA (Sanicle).—*Flowers* in panicled tufts, the outer without stamens, the inner without pistils ; *fruit* egg-shaped, covered with hooked prickles. (Name from the Latin, *sano*, to heal, the plant being formerly supposed to have remarkable healing qualities.)

3. ERYNGIUM (Eryngo).—*Flowers* in a dense prickly head ; *fruit* egg-shaped, covered with chaffy scales.

** *Umbels compound ; fruit of two flattened lobes, which are united by the narrow edge, not prickly, nor beaked.*

4. CONÍUM (Hemlock).—*Fruit* egg-shaped, each carpel with 5 wavy ridges ; *general bracts* few, *partial* 3, all on the outside. (Name, the Greek for the plant).

5. SMYRNIUM (Alexanders).—*Fruit* of two kidney-shaped carpels, each having 3 prominent ridges ; *bracts* 0. (Name from the Greek, *smyrna*, myrrh, from the scent of some of the species.)

6. CICÚTA (Water Hemlock).—*Fruit* of 2 almost globose carpels, with 5 broad, flattened ridges ; *general bracts* 1 or 2, very narrow, often 0 ; *partial*, several, unequal. (Name from the Latin, *cicúta*, a Hemlock stalk.)

7. APIUM (Celery).—*Fruit* roundish egg-shaped, of 2 almost distinct *carpels*, each with 5 slender ridges ; *bracts* 0. (Name, the Latin name of this or some allied plant.)

8. PETROSELÍNUM (Parsley).—*Fruit* egg-shaped ; *carpels*, each with 5 slender ridges ; *general bracts* few, *partial* many. (Name from the Greek, *petros*, a rock, and *selínon*, parsley.)

9. HELOSCIADIUM (Marsh-wort).—*Fruit* egg-shaped, or oblong ; *carpels*, each with 5 slender, prominent ridges ; *general bracts* 0, *partial* several. (Name from the Greek, *helos*, a marsh, and *skiádion*, an umbel.)

10. SISON (Stone Parsley).—*Fruit* egg-shaped ; *carpels* with 5 slender ridges ; *petals* broad, deeply notched, with an inflexed point ; *bracts*, both general and partial, several.—(Name, the Greek for some allied plant.)

11. ÆGOPODIUM (Gout-weed).—*Fruit* oblong ; *carpels* with 5 slender ridges ; *bracts* 0.—(Name in Greek signifying *goat's-foot*, from some fancied resemblance of the leaves.)

12. CARUM (Caraway).—*Fruit* oblong ; *carpels* with

5 slender ridges ; *general bracts* 0, or rarely 1, partial 0. (Name from *Caria*, a country of Asia Minor.)

13. BUNIUM (Earth-nut). —*Fruit* oblong, crowned with the conical base of the erect *styles ; carpels* with 5 slender, blunt ridges ; *general bracts* 0, *partial* few. (Name from the Greek, *bounos*, a hill, where the plant delights to grow.)

14. PIMPINELLA (Burnet Saxifrage).—*Fruit* oblong, crowned with the swollen base of the reflexed styles ; *carpels* with 5 slender ridges, and furrows between ; *general bracts* 0, or rarely 1, *partial* 0. (Name of doubtful etymology.)

15. SIUM (Water Parsnep).—*Fruit* nearly globose ; *carpels* with 5 slender, blunt ridges ; *bracts, general* and *partial*, several. ("Name, according to Théis, from the Celtic word *siw*, water," *Sir W. J. Hooker.*)

16. BUPLEURUM (Hare's-ear).—*Fruit* oblong ; *carpels* with 5 prominent ridges, crowned with the flat base of the styles ; *partial bracts*, very large. (Name from the Greek, *bous*, an ox, and *pleuron*, a rib, from the ribbed leaves of some species.)

*** *Umbels compound ; fruit not flattened, not prickly, nor beaked.*

17. ŒNÁNTHÉ (Water Dropwort).—*Fruit* egg-shaped, cylindrical, crowned with the long straight *styles ; carpels* with 5 blunt, corky ridges ; *flowers* somewhat rayed, those of the centre only being fertile. (Name from the Greek, *oinos*, wine, and *anthos*, a flower, from the wine-like smell of the flowers.)

18. ÆTHÚSA (Fool's Parsley).—*Fruit* nearly globose ; *carpels* with 5 sharply-keeled ridges, crowned with the reflexed *styles ; partial bracts* 3, all on one side, drooping. (Name from the Greek, *aitho*, to burn, from its acrid properties.)

19. FŒNÍCULUM (Fennel).—*Fruit* elliptical ; *carpels* with 5 bluntly-keeled ridges ; *bracts* 0. (Name from

the Latin, *fœnum*, hay, to which it has been compared
in smell.)

20. Ligústicum (Lovage).—*Fruit* elliptical ; *carpels*
with 5 sharp, somewhat winged ridges ; *bracts*, both
general and partial, several. (Name from *Liguria*, where
the cultivated species abounds.)

21. Siláus (Pepper Saxifrage).—*Fruit* egg-shaped ;
carpels with 5 sharp, somewhat winged ridges ; *petals*
scarcely notched (yellow) ; *general bracts* 1 or 2, *partial*
several. (Name given by the Romans to some probably
allied plant.)

22. Méum (Spignel).—*Fruit* elliptical ; *carpels* with
5 sharp, winged ridges ; *petals* tapering at both ends;
general bracts few, *partial* numerous. (Name, the Greek
for this or some allied plant.)

23. Crithmum (Samphire).—*Fruit* elliptical ; *carpels*
spongy, with 5 sharp, winged ridges ; *bracts*, both
general and *partial*, numerous. (Name from the Greek,
crithé, barley, to which grain the fruit bears a fancied
resemblance.)

**** *Umbels compound ; fruit of two flattened carpels,
which are united by their faces, not prickly, nor
beaked.*

24. Angélica.—*Fruit* with three sharp ridges at the
back of each *carpel*, and two at the sides, the latter
expanding into an even border ; *general bracts* few, or
0, *partial* numerous. (Named *angelic*, from its medicinal properties.)

25. Pastináca (Parsnep).—*Fruit* very flat, with a
broad border ; *carpels* with 3 slender ridges on the
back, and 2 near the outer edge of the margin ; *general*
and *partial bracts* rarely more than 1 ; (flowers yellow.)
(Name from the Latin, *pastus*, pasture.)

26. Heracléum (Cow Parsnep).—*Fruit* nearly the
same as in *Pastináca* ; *flowers* rayed ; *general bracts*
several, soon falling off, *partial* numerous. (Name from

Hercules, who is said to have brought this, or some allied plant, into use.)

***** *Umbels compound ; fruit prickly, not beaked.*

27. DAUCUS (Carrot).—*Fruit* slightly flattened ; *carpels* united by their faces, oblong ; *ridges* bristly, with a row of prickles between ; *general bracts* very long, often pinnatifid. (Name, the Greek name of the plant.)

28. CAÚCALIS (Bur-Parsley).—*Fruit* slightly flattened ; *carpels* united by thin narrow edges ; *ridges* bristly, with 1—3 rows of hooked prickles between. (Name, the Greek name of the plant.)

29. TORÍLIS (Hedge Parsley).—*Fruit* slightly contracted at the sides ; *ridges* of the *carpels* bristly, with numerous prickles between ; *partial bracts* numerous. (Name of doubtful origin.)

****** *Umbels compound ; fruit more or less beaked.*

30. SCANDIX (Shepherd's Needle).—*Fruit* contracted at the sides, with a very long beak ; *carpels* with 5 blunt ridges ; *general bracts* 0, *partial* several, longer than the flowers. (Name, the Greek name of the plant.)

31. ANTHRISCUS (Beaked Parsley).—*Fruit* with a short beak ; *carpels* without ridges ; *general bracts* 0, *partial* several. (Name, the Greek name of this or some allied plant.)

32. CHÆROPHYLLUM (Chervil).—*Fruit* contracted at the sides, with a short beak ; *carpels* with 5 blunt ridges ; *partial bracts* several. (Name in Greek signifying *pleasant leaf,* from the agreeable perfume of some species.)

33. MYRRHIS (Cicely).—*Fruit* contracted at the sides, with a deep furrow between the carpels ; *carpels* with 5 sharply-keeled ridges ; *general bracts* 0, *partial* several. (Name from the Greek, *myrrha,* myrrh, from the fragrance of the leaves.)

1. Hydrocótylé (*White-rot*).

1. *H. vulgáris* (Common White-rot, Marsh Penny-wort).—The only British species; common in marshes and bogs. A small creeping plant, very unlike the rest of the umbelliferous tribe, with round, smooth, crenate *leaves*, 1—1½ inch across, and inconspicuous heads of about 5, minute, reddish-white *flowers*, which never rise above the leaves, and require a close search to be detected at all. Each leaf is attached by its centre to the stalk, and resembles a little platter.—Fl. May, June. Perennial.

HYDROCOTYLE VULGARIS (*Common White-rot, Marsh Pennywort*).

2. Sanícula (*Sanicle*).

1. *S. Europæa* (Wood Sanicle).—The only British species; common in woods. A slender, smooth plant,

about 1½ foot high, with glossy *leaves*, which are 3—5-lobed, and cut. The *flowers* are dull white, and grow in panicled heads, rather than umbels, and are succeeded by roundish *seeds*, which are covered with hooked prickles.—Fl. June, July. Perennial.

SANÍCULA EUROPÆA (*Wood Sanicle*).

3. ERYNGIUM (*Eryngo*).

1. *E. marítimum* (Sea Eryngo, Sea Holly).—A stout prickly plant, with more of the habit of a Thistle than

one of the Umbelliferous tribe. The whole plant is remarkably rigid and glaucous. The *flowers* are blue, and grow in dense heads. The *roots* are large, fleshy, and brittle, and extend for a distance of many feet into the sand. When candied they form a well-known sweetmeat.—Fl. July, August. Perennial.

A second British species, *E. campestre* (Field Eryngo), grows at Stonehouse, Devonshire, and in two or three other places, but is very rare. It resembles the last in habit, but is taller and more slender.

ERYNGIUM MARITIMUM (*Sea Eryngo, Sea Holly*).

CONIUM MACULATUM (*Common Hemlock*).

4. Coníum (*Hemlock*).

1. *C. maculatum* (Common Hemlock).—A tall, much branched and gracefully growing plant, with elegantly cut foliage, and white flowers. Country people are in the habit of calling by the name of *Hemlock* many

species of umbelliferous plants. The real Hemlock may, however, be accurately distinguished by its slender growth, perfectly smooth *stem*, which is spotted with red ; by its finely divided *leaves*, which are also smooth, and by the *bracts* at the base of the partial *umbels*, which only go half way round. It usually grows from 2 to 4 feet high, but in sheltered situations it sometimes attains more than double that height.—Fl. June, July Biennial.

SMYRNIUM OLUSATRUM (*Common Alexanders*).

5. SMÝRNIUM (*Alexanders*).

1. *S. Olusátrum* (Common Alexanders).—A tall and stout plant, growing in waste ground, especially near

the sea ; well distinguished from any other plant of the tribe, by its broad, bright green, glossy *leaves*, which grow in threes, and by its numerous large *umbels* of greenish yellow *flowers.* The *stem* is smooth, 3—4 feet high, furrowed and hollow. The *seeds* are nearly black when ripe. The young shoots are sometimes boiled and eaten.—Fl. May, June. Biennial.

6. Cicúta (*Cowbane*).

1. *C. virósa* (Cowbane, Water-Hemlock).— Ponds and ditches ; rare. A poisonous, aquatic species ; distinguished by its very stout, hollow *stem*, pinnate and long-stalked lower *leaves*, twice ternate upper *leaves*, and stalked *umbels* of white *flowers.* The name *Water-Hemlock* is often applied to several species of *Œnanthé*, which are also very poisonous. — Fl. July, August. Perennial.

7. Apium (*Celery*).

1. *A. gravéolens* (Celery, Smallage).—Moist places near the sea. The origin of the garden Celery, and unmistakeably distinguished by its strong flavour and odour, which in no respect differ from those of the garden plant. The *stem* is usually 1—2 feet high, branched, and leafy, but sometimes nearly prostrate. The *flowers* are small and white, and grow either in terminal or axillary *umbels*, which are often sessile and unequal. In its wild state the plant is not eatable, but when it has been cultivated on rich soil, and the leaf-stalks have been blanched by being deprived of light,

s

it is a wholesome vegetable.—Fl. June—September. Biennial.

APIUM GRAVÉOLENS (*Celery, Smallage*).

8. Petroselínum (*Parsley*).

1. *P. ségetum* (Corn Parsley).—Corn fields and waste places ; not uncommon. Well distinguished by its

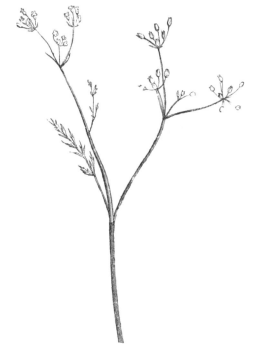

PETROSELINUM SEGETUM (*Corn Parsley*).

slender, branched *stem*, which is remarkably tough and wiry, by its small pinnated *leaves*, and *umbels* of small whitish *flowers*, and by the *rays* of the umbel being few and very unequal in length. The root-leaves wither early, and the few which grow on the stem are small and inconspicuous.—Fl. August, September.   Biennial.

* *P. sativum* is the common Parsley of gardens, which, though often found seemingly wild, is not really indigenous.

HELOSCIADIUM NODIFLORUM (*Procumbent Marsh-wort*).

9. HELOSCIADIUM (*Marsh-wort*).

1. *H. nodiflórum* (Procumbent Marsh-wort).—*Stem* prostrate and rooting ; *leaves* pinnate ; *leaflets* egg-shaped, serrated ; *umbels* on very short stalks, opposite the leaves.—In ditches and rivulets ; abundant.   A

plant with somewhat of the habit of Water-cresses, in company with which it often grows, and for which it is sometimes mistaken. It may be distinguished, when out of flower, by its serrated and somewhat pointed *leaves,* and by its hollow *stems.* The flowers are small and white.—Fl. July, Aug. Perennial.

 * *H. repens* is a smaller plant, and has narrower *leaves,* but is scarcely a distinct species : *H. inundatum* (Least Marsh-wort) has the lower *leaves* finely divided into numerous hair-like segments. The *umbels* usually have only two rays of small white *flowers,* and, with the upper leaves, are the only parts of the plant which rise out of the water.

10. SISON (*Stone Parsley*).

1. *S. Amómum* (Hedge Stone Parsley).—The only species.—Damp chalky places. A slender plant, 2—3 feet high, with a wiry, branched *stem,* and pinnate, cut *leaves,* the *leaflets* of the upper ones being very narrow. The general *umbels* consist of about 4 *rays,* with 2—4 *bracts* at the base ; the *partial umbels* are small, and have 4 *bracts* at the base of each ; the flowers are cream-coloured and very small. The whole plant has a nauseous smell.—Fl. August. Biennial.

11. ÆGOPODIUM (*Gout-weed*).

1. *Æ. podagraria* (Common Gout-weed).—A common and very troublesome garden weed, with a creeping *root,* large, thrice ternate *leaves,* and white *flowers.* The *stems* grow about a foot high. The leaves are sometimes boiled and eaten, but have a strong and very disagreeable flavour.—Fl. May, June. Perennial.

12. CARUM (*Caraway*).

1. *C. Carui* (Common Caraway).—A doubtful native, but occasionally growing wild in several parts of Great

Britain. The *stem* is much branched, about 2 feet high, the *leaves* twice pinnate, with leaflets cut into very narrow segments; the *flowers* are white, and grow in

ÆGOPODIUM PODAGRARIA (*Common Gout-weed*).

rather large umbels, with rarely more than one *bract*, and that at the base of the general umbel. The seeds are too well known to need any description.—Fl. June. Biennial.

   * *C. verticillatum* (Whorled Caraway) is a smaller plant with pinnate *leaves*, the *leaflets* of which are divided to the base into very numerous hair-like segments, and are so crowded as to appear whorled. Very rare, except in the west of Scotland.

BUNIUM FLEXUOSUM (*Common Pig-nut*).

13. Búnium (*Pig-nut*).

1. *B. flexuosum* (Common Pig-nut).—A very slender plant, about a foot high, bearing a few finely divided *leaves*, and terminal umbels of white *flowers*. The root, which resembles a small potato in shape, and is covered with a thin skin easily removed, is eatable, but only fit

for the food of the animal after which it is named.—
Fl. May, June. Perennial.

B. Bulbocástanum has been found only in chalky
fields in Cambridgeshire and Hertfordshire. It differs
from the last in having the *fruit* crowned with the
reflexed *styles*.

PIMPINELLA SAXIFRAGA (*Common Burnet Saxifrage*).

14. PIMPINELLA (*Burnet Saxifrage*).

1. *P. Saxifraga* (Common Burnet Saxifrage). — A
slender plant, 1—2 feet high, common in dry pastures.
The lower leaves, which are pinnate, with sharply cut
leaflets, grow on long stalks ; the upper ones twice pinnate, and deeply cut into very narrow, sharp segments.—
Fl. July, August. Perennial.

P. magna (Greater Burnet Saxifrage) is stouter and

larger than the last, and has all the *leaves* pinnate, the terminal *leaflet* on each being 3-lobed.   It grows in shady places, but is far from common.

15. Sıum (*Water Parsnep*).

1. *S. latifolium* (Broad-leaved Water Parsnep).— *Leaves* pinnate ; *leaflets* narrow oblong, pointed, equally

serrated; *umbels* terminal; *bracts*, both *general* and *partial*, narrow and pointed.—Watery places; not common. A stout plant, with a furrowed stem, 3—5 feet high, pinnated leaves of 5—13 large and distant leaflets, and long, flat umbels of white flowers.—Fl. July, August. Perennial.

2. *S. angustifolium* (Narrow-leaved Water Parsnep). *Leaves* pinnate; *leaflets* unequally cut, egg-shaped, the upper ones narrower; *umbels* opposite the leaves, stalked.—Watery places; not unfrequent. Smaller than the last, and resembling *Helosciádium nodiflórum*, from which it may be distinguished by its stalked *umbels*, and by its having general and partial *bracts*, which are reflexed and often cut.

16. BUPLEURUM (*Thorow-wax*).

1. *B. rotundifolium* (Common Thorow-wax, Hare's-ear). — *Stem* branched above; *leaves* roundish egg-shaped, undivided, perfoliate; *general bracts* wanting, *partial* ones large, bristle pointed, thrice as long as the flowers.— Corn-fields, on a chalky soil. A singular plant, well distinguished by its perfoliate leaves, which have a glaucous hue, and by its large, greenish yellow, partial bracts, which are far more conspicuous than the minute yellow flowers.—Fl. July. Annual.

2. *B. tenuissimum* (Slender Hare's-ear) is remarkable for its slender, wiry *stem*, about a foot high, its very narrow undivided *leaves*, and small *umbels* of very few, minute, yellowish *flowers*; it grows in salt marshes on the south and east coasts of England: *B. aristatum* (Narrow-leaved Hare's-ear) is a small plant, 3—4 inches high, with pale, rigid *leaves*, inconspicuous greenish *flowers*, and large, sharp-pointed *bracts*. It is found nowhere in Great Britain but on the cliffs at Torquay, Devon.

17. Œnanthé (*Water Dropwort*).

1. *Œ. fistulosa* (Tubular Water Dropwort).—*Root* sending out runners; *stem-leaves* pinnate, shorter than

BUPLEURUM ROTUNDIFOLIUM (*Common Thorow-wax, Hare's-ear*).

their tubular stalks.—Watery places; not uncommon. Well marked by its tubular stems, leaves, and leaflets. The lower leaves are entirely submersed, and of these the leaflets are flat, but all the rest of the plant consists of a series of tubes. The *umbels* are of very few rays, which when in fruit are nearly globular.—Fl. July, August. Perennial

ŒNANTHE CROCATA (*Hemlock Water Dropwort*).

2. *Œ. crocáta* (Hemlock Water Dropwort.)—*Leaves* thrice pinnate ; *leaflets* wedge-shaped, variously cut.— Watery places, common.—A large and stout plant, 3—5 feet high, with clustered, tuberous roots, somewhat like those of the Dahlia, spreading, glossy leaves, and large umbels of white flowers. The plant is popularly known by the name of *Water Hemlock,* and, being very poison-ous, should not be allowed to grow in places where cattle

are kept, as instances are numerous in which cows have been poisoned by eating the roots.

* *Œ. Phellandrium* differs from the preceding in having the *leaves* divided into very fine segments, and fibrous, not tuberous, *roots*. *Œ. pimpinelloídes* is a much slenderer plant, with very narrow *leaflets* and small white *flowers* crowded into round *heads;* it grows in salt marshes. *Œ. peucedanifólia* resembles the last, but grows in fresh water.

ÆTHUSA CYNAPIUM (*Fool's Parsley*).

18. ÆTHÚSA (*Fool's Parsley*).

1. *Æ. Cynapium* (Fool's Parsley).—A slender plant about a foot high, with dark green, doubly pinnate *leaves,*

and terminal *umbels* of white *flowers*. It is a common garden weed, and in its young state somewhat resembles Parsley, but when in flower may readily be distinguished from that and all other umbelliferous plants, by having no *general bracts*, but at the base of each *partial umbel* three very long and narrow *leaves*, which are all on the outer side, and point downwards. It is said, and probably with reason, to be poisonous.—Fl. July, August. Annual.

19. Fœnículum (*Fennel*).

1. *F. vulgare* (Common Fennel).—Waste places, especially near the sea, common. A well-known plant, with an erect rod-like *stem*, numerous *leaves*, which are deeply divided into soft hair-like segments, and large terminal *umbels* of yellow *flowers*. The whole plant is aromatic, and the chopped leaves are often used as an ingredient in sauce for fish.—Fl. July, August. Perennial.

20. Ligústicum (*Lovage*).

1. *L. Scóticum* (Scottish Lovage).—Rocky sea-shore in Scotland and Northumberland. About 1½ foot high, *stem* nearly simple, tinged with red ; *leaves* twice ternate, with large broad serrated leaflets, *umbels* with general and partial *bracts ; flowers* reddish-white.—Fl. July. Perennial.

21. Siláus (*Pepper Saxifrage*).

1. *S. pratensis* (Meadow Pepper Saxifrage).—Meadows, not very general. From 1 to 2 feet high ; *leaves* thrice pinnate, with narrow opposite leaflets, and terminal *umbels* of dull yellowish-white *flowers ; general bracts*

FŒNICULUM VULGARE (*Common Fennel*).

1—3, *partial*, numerous.—" The whole plant being fetid when bruised, is supposed in some parts of Norfolk to give a bad flavour to milk and butter; but cattle do not eat it, except accidentally, or in small quantities, though sufficient perhaps to have the effect in question."—*Sir J. E. Smith.*—Fl. July, September.   Perennial.

MEUM ATHAMANTICUM (*Spignel, Meu, or Bald-Money*).

22. MÉUM (*Spignel*).

1. *M. athamánticum* (Spignel, Meu, *or* Bald-Money).
—Dry mountainous pastures in the north. Well dis-

'tinguished by its twice pinnate *leaves*, the *leaflets* of which are divided into numerous thread-like segments. The whole plant, and especially the root, which is eaten by the Highlanders, is highly aromatic, with a flavour like Melilot, which it communicates to milk and butter, from the cows feeding on its herbage in spring.—"Bald, or Bald-money, is a corruption of *Balder*, the *Apollo* of the northern nations, to whom this plant was dedicated." —*Sir W. J. Hooker.*—Fl. June, July. Perennial.

CRITHMUM MARITIMUM (*Sea Samphire*).

23. CRITHMUM (*Samphire*).

1. *C. maritimum* (Sea Samphire).—Rocks by the sea-side, abundant. Well distinguished by its long, glaucous, fleshy *leaflets*, and yellow *flowers*. The whole plant is aromatic, and has a powerful scent. The young leaves,

T

if gathered in May, sprinkled with salt, and preserved in vinegar, make one of the best of pickles. On those parts of the coast where Samphire does not abound, other plants, which resemble it in having fleshy leaves, are sometimes sold under the same name, but are very inferior.—Fl. July, August. Perennial.

ANGELICA SYLVESTRIS (*Wild Angélica*).

24. ANGÉLICA.

1. *A. sylvestris* (Wild Angélica).—Wet places, common. A stout and tall plant, 2—4 feet high ; the *stem* is furrowed, tinged with purple, and slightly downy, especially in its upper part ; the *leaves* are twice pinnate ; the *leaflets* egg-shaped and serrated, the *umbels*

are large, and furnished with both *general* and *partial bracts;* the *flowers* are white, tinged with pink.—Fl. July. Perennial.

* *A. Archangélica* is a larger species, commonly cultivated for the sake of its aromatic stems, which when candied form a favourite sweetmeat.

PASTINACA SATIVA (*Common Parsnep*).

25. Pastináca (*Parsnep*).

1. *P. sativa* (Common Parsnep).—Well known in gardens as an agreeable and nutritive vegetable. In its wild state, the plant, which is not uncommon in lime-

stone and chalky pastures, closely resembles the cultivated variety, but has smaller roots and more downy
leaves. The *flowers* are yellow, and grow in terminal
umbels.—Fl. July, August. Biennial.

HERACLEUM SPONDYLIUM (*Common Cow-Parsnep, Hog-weed*).

26. HERÁCLEUM (*Cow-Parsnep*).

1. *H. Spondýlium* (Common Cow-Parsnep, Hog-weed).
—A very tall and stout plant, with a channeled, hairy

stem, 4—6 feet high, large, irregularly cut, rough *leaves*, and spreading *umbels* of conspicuous white *flowers*. In spring the plant is remarkable for the large oval tufts formed by the sheathing base of the stem leaves, which contain the flower buds. This, with many other umbelliferous plants, is often confounded by farmers with Hemlock, and great pains are taken to eradicate it; but cattle eat it with impunity, and it is probably a wholesome and nutritive food. It is often very abundant in meadows.—Fl. July. Biennial.

27. Daucus (*Carrot*).

1. *D. Caróta* (Wild Carrot). — A tough-stemmed, bristly plant, with much-cut *leaves*, and large *umbels* of dull white *flowers*. Well distinguished by having the central *flower* or *partial umbel* of flowers bright red or deep purple. In flavour and scent it resembles the garden carrot.—Fl. July, August. Biennial.

   * A variety (*D. marítimus*), abundant on many parts of the sea-coast, differs from the preceding in having somewhat fleshy leaves, and in being destitute of the central purple flower, or umbel.

28. Caúcalis (*Bur-Parsley*).

1. *C. daucoídes* (Small Bur-Parsley). — *Leaves* repeatedly divided; *umbels* of about 3 rays, without bracts; *partial umbels* of few flowers, with about 3 bracts.— Chalky fields, not common. A somewhat bushy plant, nearly smooth, with a stem which is deeply furrowed and hairy at the joints. The flowers, which are pinkish white, grow both in lateral and terminal umbels, and are succeeded by large prickly seeds.—Fl. June. Annual.

DAUCUS CAROTA (*Wild Carrot*).

* *C. latifólia* (Great Bur-Parsley), is not uncommon in chalky fields in Cambridgeshire. It is taller than the preceding, and is well distinguished from it and all other British plants of the tribe by its handsome, large, rose-coloured flowers.

TORILIS ANTHRISCUS (*Upright Hedge-Parsley*).

29. Torílis (*Hedge-Parsley*).

1. *T. Anthriscus* (Upright Hedge-Parsley).—*Leaves* twice pinnate; *leaflets* narrow, sharply cut; *umbels* stalked; *general* and *partial bracts* several.—Hedges, abundant. A tall slender plant 2—3 feet high, with a solid rough stem, hairy leaves, and many rayed umbels of small white or pinkish flowers. The fruit is thickly

covered with incurved rigid bristles.—Fl. July, August. Annual.

2. *T. infesta* (Spreading Hedge-Parsley). — *Leaves* twice pinnate; *leaflets* oblong, sharply cut; *umbels* stalked; *general bracts* 1 or 0, *partial* several.—Hedges, common. Smaller than the last, 6—18 inches high, with more branched stems and more rigid leaves. The fruit is covered with spreading hooked bristles.—Fl. July, August. Annual.

3. *T. nodósa* (Knotted Hedge-Parsley).—*Stem* prostrate; *umbels* simple, lateral, nearly sessile.—Hedges and waste places, common. Well distinguished from all other British umbelliferous plants by its prostrate mode of growth, its very small, almost globular umbels of whitish flowers, and by the outer carpels in each umbel being covered with hooked prickles, while the inner are warty.—Fl. May, July. Annual.

30. SCANDIX (*Shepherd's Needle*).

1. *S. Pecten* (Shepherd's Needle, Venus' Comb).— Common in cultivated ground. A small plant, 3—9 inches high, with finely-cut bright-green *leaves*, and few-rayed umbels of small white *flowers*, which are succeeded by long, beaked *seed-vessels.*—Fl. June, September. Annual.

31. ANTHRISCUS (*Beaked-Parsley*).

1. *A. vulgaris* (Common Beaked-Parsley). — *Stem* smooth; *leaves* thrice pinnate, with blunt segments; *umbels* lateral on rather short stalks; *fruit* bristly.— Waste ground, chiefly near towns. Remarkable for its smooth, polished stem and delicate green leaves, which are slightly hairy beneath. The stem is 2—3 feet high, slightly swollen under each joint. The flowers are white,

and grow in umbels opposite the leaves, without general bracts ; partial bracts 5 or 6, with fringed edges.—Fl. May. Annual.

SCANDIX PECTEN (*Shepherd's Needle, Venus' Comb*).

2. *A. sylvestris* (Wild Beaked-Parsley).—*Stem* slightly downy below, smooth above ; *leaves* thrice pinnate, the segments rough-edged ; *umbels* terminal on long stalks, drooping when young ; *fruit* smooth.—Hedges, common. One of our early spring flowers, distinguished, when in bud, by the drooping partial umbels, each of which has about 5 reflexed bracts ; and afterwards, by its smooth, shortly-beaked fruit.—Fl. April, June. Perennial.

ANTHRISCUS SYLVESTRIS (*Wild Beaked-Parsley*).

* *A. Cerefolium* (Garden Chervil), is not a native plant, though sometimes found in the neighbourhood of houses. It may be distinguished from the preceding by having only 3 partial *bracts*, lateral *umbels*, and smooth *fruit*.

32. CHÆROPHYLLUM (*Chervil*).

1. *C. temulentum* (Rough Chervil).—The only British species, very common in woods and hedges. The *stem* is slender, 2—3 feet high, rough with short hairs, spotted

with purple, and swollen beneath the joints; the *leaves* are twice pinnate, deeply lobed and cut, hairy, often making the plant conspicuous in autumn by their rich purple hue; the *flowers* are white, and grow in terminal *umbels*, which droop when in bud; *general bracts* either absent or very few, *partial bracts* several, fringed and deflexed. —Fl. June, July. Perennial.

33. MYRRHIS (*Cicely*).

1. *M. odoráta* (Sweet Cicely).—Mountainous pastures in the North. Remarkable for its sweet and highly aromatic flavour. The *stem* is 2—3 feet high, furrowed and hollow; the *leaves* are large, thrice pinnate, cut, and slightly downy. The *flowers* are white, and grow in terminal downy umbels; *bracts*, partial only, whitish and finely fringed. The *fruit* is remarkably large, an inch long, dark brown, with very sharp *ribs*, and possesses the flavour of the rest of the plant in a high degree.—Fl. May, June. Perennial.

The foregoing descriptions contain only those umbelliferous plants which are most commonly to be met with. There are besides these a few others, which are either of unusual occurrence, or have escaped from cultivation; these are:—

Physospernum Cornubiense (Cornish Bladder-seed).— An erect, smooth plant, about 2 feet high, with thrice ternate *leaves* and white *flowers*, which are furnished with both *general* and *partial bracts*. The *fruit* when ripe is remarkably inflated and nearly globose, whence its name. It is found only near Bodmin, Cornwall, and Tavistock, Devon.

Trínia glabérrima (Honewort), grows on limestone rocks in Somersetshire, and at Berry-head, Devon. It may be distinguished from all other British umbelliferous plants by bearing its *stamens* and *pistils* in separate flowers and on different plants.

Séseli Libanótis (Mountain Meadow-Saxifrage) is of rare occurrence, in Cambridgeshire and Sussex. It may be distinguished by its hemispherical *umbels* and hairy *fruit*, covered with the reflexed *styles*.

Peucédanum officinále (Sea Hog's Fennel) a rare plant, remarkable for its large umbels of yellow *flowers*, occurs in salt marshes on the eastern coast of England. *P. palustre* (Marsh Hog's Fennel) is also a rare species, growing in marshes in Yorkshire, Lancashire, &c. The *stem* grows 4—5 feet high, and abounds in a milky juice, which dries to a brown resin.

Coriandrum sativum (Common Coriander) is occasionally found in the neighbourhood of towns, but cannot be deemed a native plant. It is well marked by its globose, pleasantly aromatic *fruit*.

ORD. XXXIX.—ARALIACEÆ.—THE IVY TRIBE.

Calyx attached to the ovary, 4—5-cleft ; *petals* 4, 5, or 10, occasionally wanting; *stamens* equalling the petals in number, or twice as many, inserted on the ovary ; *ovary* with more than 2 cells ; *styles* as many as the cells ; *fruit* fleshy or dry, of several 1-seeded cells.— Trees, shrubs, or herbaceous plants, not confined to any particular climate, closely resembling the Umbelliferous Tribe in the structure of their flowers, but not partaking their dangerous properties. Only two species are natives of Britain ; but one of these, Ivy, is so universally diffused, as to be familiar to every one ; the other, Moschatell, is a humble plant, with solitary heads of green flowers, and delicate leaves strongly scented with musk. *Ginseng*, the favourite medicine of the Chinese, is the root of some species of *Panax*, a plant belonging to this tribe. Some species of Ivy furnish wood scented like Lavender and Rosemary, and others an aromatic gum. A remarkable plant belonging to this order is *Gunnera scabra*, found by Darwin growing on the sandstone cliffs of Chiloe. He describes it as somewhat resembling Rhubarb on a gigantic scale, each plant producing four or five leaves nearly eight feet in diameter.

1. HEDERA (Ivy).—*Calyx* of 5 teeth, inserted in the ovary; *petals* 5—10 ; *stamens* 5—10; *styles* 5—10, often combined into 1 ; *berry* 5-celled and 5-seeded, crowned by the calyx. (Name, the Latin name of the plant.)

2. ADOXA (Moschatell).—*Calyx* 3-cleft, inserted above the base of the ovary ; *corolla* 4 or 5-cleft, inserted on the ovary; *stamens* 8 or 10, in pairs; *anthers* 1-celled ; *berry* 4 or 5-celled. (Name, in Greek signifying *inconspicuous*, from its humble growth.)

HEDERA HELIX (*Common Ivy*).

1. Hédera (*Ivy*).

1. *H. Helix* (Common Ivy).—The only British species, too well known to need any description. (*See* " Forest Trees of Britain," vol. ii.)—Fl. Oct. Nov.   Shrub.

ADOXA MOSCHATELLINA (*Common Moschatell*).

2. ADOXA (*Moschatell*).

1. *A. Moschatéllina* (Common Moschatell). — The only species. A small herbaceous plant 4—6 inches high, growing in damp woods and hedge banks, and not unfrequent at a great elevation among the mountains. Each plant bears several delicate root-leaves, and two smaller leaves half-way up the stem. The *flowers* grow in terminal heads of 5 each, the upper flower with 4 *petals* and 8 *stamens*, the four side flowers having 5 *petals* and 10 *stamens* each; the latter are remarkable for being inserted in pairs, and for bearing 1-celled *anthers;* or the filaments may be considered to be forked, each fork bearing the lobe of an anther. The whole plant diffuses a musk-like scent, which, however, is not perceptible if the plant be bruised.—Fl. April, May. Perennial.

ORD. XL.—CORNEÆ.—THE CORNEL TRIBE.

Sepals 4, attached to the ovary; *petals* 4, oblong, broad at the base, inserted into the top of the calyx; *stamens* 4, inserted with the petals; *ovary* 2-celled; *style* thread-like; *stigma* simple; *fruit* a berry-like drupe, with a 2-celled nut; *seeds* solitary.—Mostly trees or shrubs, with opposite leaves, and flowers growing in heads or umbels. A small order, containing few plants of interest, which inhabit the temperate regions of Europe, Asia, and America. In the United States several species are found, the bark of which is a powerful tonic, ranking in utility next to Peruvian bark. *Benthamia fragifera*, a handsome shrub from the mountains of Nepal, was introduced into England in 1825. In Cornwall, where it was first raised from seed, it flowers and bears fruit freely, and forms a pleasing addition to the shrubbery. Two species of *Cornus* are indigenous to Britain. The *Cornus* of the ancients was

the present Cornelian Cherry, *Cornus máscula,* whose little
clusters of yellow starry flowers are among the earliest
heralds of spring. Its fruit is like a small plum, with
a very austere flesh, but after keeping it becomes pleasantly acid. The Turks still use it in the manufacture
of sherbet. A similar species is commonly cultivated in
Japan for the sake of its fruit, which is a constant ingredient in the acid drinks of that country. The shrub
now common in gardens under the name of Spotted
Laurel (*Aúcuba Japonica*) belongs to this order.

CORNUS SANGUINEA (*Wild Cornel,* Dog-*wood*).

1. CORNUS (Cornel).—Characters described above. (Name, from the shrub so called by the Latins, from the horn-like nature of its wood.)

1. CORNUS (*Cornel*).

1. *C. sanguinea* (Wild Cornel, Dog-wood).—Shrubby. —Hedges and thickets, especially on a chalk or lime-stone soil. A bushy shrub 5—6 feet high, with oppo-

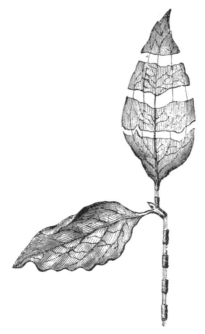

LEAF OF THE WILD CORNEL.

site, egg-shaped, pointed *leaves,* and terminal *cymes* of dull-white flowers. The bark on the old branches is of a reddish hue ; the berries are small, and dark purple. The young shoots in spring, and the leaves at all seasons, are particularly well adapted for displaying the presence of spiral vessels in these parts, being numerous and remarkably strong. The wood is much sought after by butchers for making skewers. The name Dog-wood is often given to the Spindle Tree (*Euónymus Europœa*), and Guelder Rose (*Viburnum Lantána*), trees which have wood of similar texture —Fl. June. Shrub.

2. *C. Suécica* (Dwarf Cornel).—Herbaceous.—Mountainous pastures in Scotland and the north of England. Very different in habit from the last ; *root* woody, creeping, and sending up annual flowering stems, which are about 6 inches high, and bear each an *umbel* of dark purple *flowers* with yellow *stamens :* at the base of each umbel are 4 egg-shaped yellow *bracts* tinged with purple. The fruit, which is red, is said by the Highlanders to create appetite, and hence is called *Lus-a-chraois*, plant of gluttony.—Fl. July, August. Perennial.

<hr />

Sub-class III.

COROLLIFLORÆ.

Petals united, bearing the *stamens*.

Ord. XLI.—LORANTHEÆ.—Mistletoe Tribe.

Stamens and *pistils* usually on different plants.— *Calyx* attached to the *ovary*, with 2 bracts at the base, sometimes almost wanting ; *petals* 4—8, united at the base, expanding in a valve-like manner ; *stamens* equalling the petals in number, and opposite to them ; *ovary* 1-celled ; *style* 1 or 0 ; *stigma* simple ; *fruit* succulent, 1-celled, 1-seeded ; *seed* germinating only when attached to some growing plant of a different species.—**Shrubby**

plants of singular structure and habit, growing only
(with rare exceptions) on the branches of other trees,
and therefore true parasites. The leaves are usually in
pairs, fleshy, and without veins ; the flowers inconspicuous ; but this is not always the case, for one species,
Nuytsia floribunda, which grows in the neighbourhood
of King George's Sound, bears an abundance of bright
orange-coloured flowers, producing an appearance which
the colonists compare to a tree on fire, and hence they
call it the *Fire-tree*. This species is not a parasite, but
the greater part of the tribe refuse to grow except on
living vegetables. The seed of most species is coated
with a viscid substance, by which it adheres to the bark,
and which in a few days becomes a transparent glue.
Soon, a thread-like radicle is sent forth, which, from
whatever part of the seed it proceeds, curves towards
the supporting tree, and becomes flattened at the extremity like the proboscis of a fly. Finally, it pierces the
bark, and roots itself in the growing wood, having the
power of selecting and appropriating to its own use such
juices as are fitted for its sustenance. Though not
themselves furnished with milky juices, they sometimes
attach themselves to plants of that description, and extract only such nutriment as they require. The wood
of the common Mistletoe has been found to contain twice
as much potash, and five times as much phosphoric acid,
as the wood of the foster tree. Great virtues were attributed to the Mistletoe by the Druids, but at present its
medicinal properties are in no repute. A species of
Loranthus is used in Chili as a dye, but with this exception there are scarcely any which possess useful properties.
About 400 species have been discovered, which inhabit
principally the tropical regions of Asia and America.

1. VISCUM (Mistletoe).—Stamens and pistils on separate plants ; *Barren flower, calyx* 0 ; *petals* 4, fleshy,
united at the base, each bearing an anther ; *Fertile
flower, calyx* a mere rim ; *petals* 4, very small ; *stigma*
sessile ; *berry* 1-seeded, crowned by the calyx. (Name

the Latin name of the plant, derived, it is said, from
the Celtic *gwid*, the shrub, as being the most sacred
of plants.)

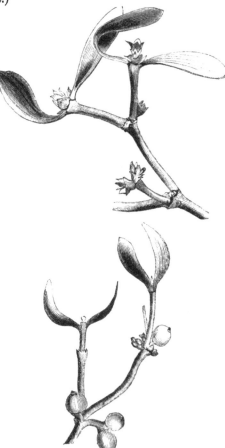

VISCUM ALBUM (*Common Mistletoe.*)

1. Viscum (*Mistletoe*).

1. *V. album* (Common Mistletoe).—The only British species. Growing on a great variety of trees especially the Apple, exceedingly rare on the Oak. The stem is green and smooth, separating easily when dead into bone-like joints; the leaves are thick and leathery, of a yellow hue, the whole plant being most conspicuous in winter, when its white berries ripen.—Fl. March—May. Perennial.

Ord. XLII.—CAPRIFOLIACEÆ.—The Woodbine Tribe.

Calyx attached to the ovary, usually with *bracts* at the base; *corolla* regular or irregular, 4—5-cleft; *stamens* equal in number to the lobes of the corolla and alternate with them; *ovary* 3—5-celled; *stigmas* 1—3; *fruit* usually fleshy, crowned by the calyx.—This tribe comprises shrubs and herbaceous plants of very different habits, and is interesting from containing the fragrant Honeysuckle or Woodbine, and the elegant little plant which Linnæus fixed on to commemorate his name. They are principally confined to the northern hemisphere, and, among them, several are natives of Britain, which will be found described at length in "Forest Trees of Britain," vol. ii. The common Elder was formerly held in high repute for its medicinal properties; and preparations of the leaves, flowers, and fruit are still used in medicine in rural districts. The roasted berries of *Triósteum perfoliatum* have been used as a substitute for coffee. *Leycestéria formósa*, a beautiful shrub from the mountains of Nepal, where it flowers at an elevation of from 6,000 to 8,000 feet, is now becoming a common ornament in our gardens. It is most attractive when in a flowering state, from the contrast of the deep green hue of its stem and leaves with the purple colour of its floral leaves and berries.

1. SAMBÚCUS (Elder).—*Calyx* 5-cleft ; *corolla* wheel-shaped, 5-lobed ; *stamens* 5 ; *stigmas* 3, sessile ; *berry* 3—4-seeded. (Name from the Greek, *sambúcé,* a musical wind-instrument, in making which the wood was anciently employed.)

2. VIBURNUM (Guelder Rose).—*Calyx* 5-cleft ; *corolla* funnel-shaped, 5-lobed ; *stamens* 5 ; *stigmas* 3, sessile ; *berry* 1-seeded. (Name, the Latin name of the plant.)

3. LONICÉRA (Honey-suckle).—*Calyx* small, 5-toothed; *corolla* tubular, irregularly 5-cleft ; *stamens* 5 ; *style* thread-shaped ; *stigma* knobbed ; *berry* 1—3-celled, with several cells. (Named in honour of Adam Lonicer, a German botanist.)

4. LINNÆA. — *Calyx* 5-cleft ; *corolla* bell-shaped, 5-cleft, regular ; *stamens* 4, two longer ; *fruit* dry ; 3-celled, one cell only containing a single seed.

1. SAMBÚCUS (*Elder*).

1. *S. nigra* (Common Elder).—*Stem* woody ; *cymes* with 5 principal branches.—A small tree, abundant in most places. Remarkable for the large quantity of pith contained in the young branches, and for the elasticity of its wood. The *leaves* are pinnate, of a strong and unpleasant odour, the flowers white and of a sickly smell, the fruit dark purple, or rarely white —Evelyn, speaking in its praise, says : " If the medicinal proper-ties of the leaves, bark, berries, &c. were thoroughly known, I cannot tell what our countrymen could ail for which he would not find a remedy from every hedge, either for sickness or wound."—Fl. June.   Tree.

* A smaller species (*S. Ébulus*), Dwarf Elder, is occa-sionally found in bushy places, and differs from the preceding in having an herbaceous *stem* 2—3 feet high, narrower *leaves,* and 3-cleft *cymes.*

SAMBUCUS NIGRA (*Common Elder*).

VIBURNUM LANTANA (*Wayfaring-tree, Mealy Guelder-rose*).

2. VIBURNUM (*Guelder Rose*).

1. *V. Lantána* (Wayfaring-tree, Mealy Guelder-rose). *Flowers* in cymes, all perfect; *leaves* elliptical, heart-shaped at the base, serrated, very downy beneath.— A large shrub, with white, mealy, flexible branches, and large leaves, which are very hoary beneath. The flowers are dull white and grow in terminal *cymes;* the berries are scarlet, turning black when fully ripe.—It is most

frequently met with in a chalky or limestone soil.—Fl.
May, June. Shrub.

VIBURNUM OPULUS (*Guelder Rose, Water Elder*).

2. *V. Ópulus* (Guelder-rose, Water Elder).—*Flowers*
in cymes, the outer barren and radiant ; *leaves* smooth,
lobed, and cut ; *leaf-stalks* with glands at the upper
extremity.—Moist woods and hedges, not uncommon.
Forming a shrub or small tree, with conspicuous snow-
white flowers ; the leaves assume a rich purple hue
before falling, when they are very ornamental, as are
also the bunches of coral-red berries, which it is said
are sometimes fermented and eaten. This statement,
however, seems scarcely credible to any one who has

chanced to smell them. The bark is very acrid. In the wild plant the cyme is flat, the outer flowers being large and showy, but destitute of stamens and pistils; in the garden variety, called the Snowball-tree, the cyme is composed entirely of barren flowers collected into a globular form.—Fl. June, July. Shrub, or small tree.

LONICERA PERICLYMENUM (*Honeysuckle, Woodbine*).

3. LONICÉRA (*Honeysuckle*).

1. *L. Periclýmenum* (Honeysuckle, Woodbine).—*Stem* twining; *flowers* gaping, in terminal heads; *leaves* all distinct (not united at the base).—A common and favourite twining shrub, the first to expand its leaves in

spring, or rather in winter, and almost the last to blossom
in autumn. Though highly ornamental to our woods,
it is decidedly injurious to young trees, clasping them
so tightly, as to mark the rind with a spiral line, and
finally becoming embedded in the wood. Handsome
twisted walking-sticks are thus formed, but the growth
of the tree is greatly checked. The leaves are not unfrequently lobed like those of the oak.—Fl. July, and
again in October. Shrub.

    * Two other species of Honeysuckle are occasionally
found, but are not considered natives of Britain. *L.
perfoliatum* (Pale perfoliate Honeysuckle), which is

LINNÆA BOREALIS (*Linnæa*).

distinguished by having the uppermost pair of *leaves* *connate*, or united by their bases ; and *L. Xylósteum* (Upright Fly-Honeysuckle), an erect shrub, with downy *leaves* and pale-yellow, scentless *flowers*, which grow in pairs.

4. LINNÆA.

L. Boreális (Linnæa).—The only species. Deservedly regarded with peculiar interest as being the " little northern plant, long overlooked, depressed, abject, flowering early," which Linnæus himself selected as therefore most appropriate to transmit his name to posterity.—It grows in woods especially of Fir, in Scotland, and in one English station, namely, a plantation of Scotch Firs in the parish of Hartburn, Northumberland. The *stem* trails along the ground, and bears at intervals pairs of roundish, slightly crenate *leaves*. The flowering *stalks* are erect, and bear each two pendulous bell-shaped *flowers*, which are fragrant, and of a delicate pink colour.—Fl. June, July. Perennial.

ORD. XLIII.—RUBIACEÆ.—THE MADDER TRIBE.

Calyx 4—6-lobed, or wanting ; *corolla* 4—6-lobed, wheel-shaped or tubular, regular ; *stamens* equal in number to the lobes of the corolla, and alternate with them ; *ovary* 2-celled ; *style* 2-cleft ; *stigmas* 2 ; *pericarp* 2-celled, 2-seeded. This order, taken in its widest extension, is one of the largest with which we are acquainted, containing more than 2,800 species, of which some are of the highest utility to man, both as food and medicine. Among the former, *Coffea Arábica* holds the first place. The seeds of this tree furnish the coffee of commerce. Several species of *Cinchóna*, a South American family, furnish Peruvian Bark and Quinine ; and drugs of similar properties are obtained from other plants of the same tribe. Ipecacuanha is the powdered root of a small plant which grows in damp, shady forests in Brazil. The wood of another plant of this tribe, *Evosmia corymbosa*, is so poisonous, that Indians have been poisoned by eating meat roasted on spits made of it. Not a few, moreover, are noted for the fragrance and beauty of their flowers. All the above-mentioned are natives of hot climates ; the British species are very different, both in habit and properties. They are herbaceous plants, with slender angular stems, whorled leaves, and small flowers, possessing no remarkable properties, except that of containing a colouring matter in their roots, which is used as a dye. This group has been separated by some botanists, and made to constitute a distinct order, under the name of STELLATÆ, a name particularly appropriate to them, from the *star-like* arrangement of their leaves. The most important of all of these is *Rúbia tinctoria*, the roots of which afford Madder, a valuable dye, and possess the singular property of imparting a red colour to the bones of animals which feed on it. Other species of Rubia, growing in Bengal and China, are used for the same purpose. No British species are of any great value, though it is said that the seeds of *Galium*, when roasted, are a good substitute for coffee, and the flowers of *Galium verum* are used as rennet to curdle milk. The most attractive British species is Woodruff, well known for the fragrance of its leaves when dry.

1. Rúbia (Madder).—*Corolla* wheel-shaped, or bell-shaped ; *stamens* 4 ; *fruit* a 2-lobed berry. (Name, from the Latin *ruber*, red, from the dye of that colour afforded by some species.)

2. Galium (Bed-straw). — *Corolla* wheel-shaped ; *stamens* 4 ; *fruit* dry, 2-lobed, 2-seeded, not crowned by the calyx. (Name, from the Greek *gala*, milk, for curdling which some species are used.)

3. Aspérula (Woodruff).—*Corolla* funnel-shaped ; *stamens* 4 ; *fruit* dry, 2-lobed, 2-seeded, not crowned by

the calyx. (Name, from the Latin *asper*, rough, from the roughness of the leaves of some species.)

4. SHERARDIA (Field-Madder). — *Corolla* funnel-shaped ; *stamens* 4 ; *fruit* dry, 2-lobed, 2-seeded, crowned by the calyx. (Named in honour of James Sherard, an eminent English botanist.)

RUBIA PEREGRINA (*Wild-Madder*).

1. RUBIA (*Madder*).

1. *R. peregrina* (Wild-Madder).—The only British species, common in bushy places in the south-west of England. A long straggling plant, many feet in length,

with remarkably rough *stems* and *leaves,* the latter growing in whorls of 4—6, glossy above, and recurved at the
margin. The *flowers* are green, 5-cleft, and grow in
panicles ; the *berries* remain attached to the plant until
late in winter ; they are black, and about as large as
currants.—Fl. June—August. Perennial.

GALIUM APARINE (*Goose-Grass*).

2. GALIUM (*Bed-straw*).

* *Flowers yellow.*

1. *G. cruciatum* (Cross-wort).—*Leaves* 4 in a whorl,
soft and downy ; *flowers* clustered in the whorls, *upper*

ones having pistils only, *lower*, stamens only.—Bushy hedges, common; well distinguished by its uniform leaves, and short clusters of dull yellow flowers. The stems are scarcely branched, 1—2 feet long.—Fl. May, June. Perennial.

2. *G. verum* (Yellow Bed-straw).—*Leaves* about 8 in a whorl, very narrow (almost thread-like), grooved and often downy below.—Dry banks, especially near the sea, where it forms a conspicuous object with its dense panicles of golden-yellow flowers. The Highlanders use the roots, in conjunction with alum, to dye red, and the rest of the plant as rennet to curdle milk.—Fl. July, August. Perennial.

** *Flowers white.*

† *Fruit smooth.*

3. *G. Mollúgo* (Hedge Bed-straw).—*Leaves* 8 in a whorl, oblong, tapering at each end, with a bristly point, roughish at the edge; *flowers* in a loose, spreading panicle.—Common everywhere in England, but said to be less frequent in Scotland. This and *G. Apariné* are the most abundant species, and resemble one another in having long straggling stems; *G. Apariné* has the prickles pointing backwards, and clings to the dress when touched, but in the present species, the prickles on the edges of the leaves are weak and point forwards. —Fl. July, August. Perennial.

4. *G. palustré* (Water Bed-straw).—*Leaves* 4—6 in a whorl, oblong, blunt, tapering at the base; *stem* weak, straggling, more or less rough; *flowers* in loose, spreading panicles.—Watery places, common. Variable in size and roughness; likely to be confounded only with the following, from which it differs in its superior size and blunt leaves, which are frequently unequal in length, especially in the upper whorls.—Fl. July, August. Perennial.

5. *G. uliginósum* (Rough Marsh Bed-straw).—*Leaves* 6—8 in a whorl, narrow, tapering at both ends, bristle-pointed, their edges as well as the angles of the *stem* rough with prickles, which point backwards.—Watery places, common. Smaller than the last, the stem being rarely more than a foot high, slender and brittle; panicles of few flowers.—Fl. July, August. Perennial.

6. *G. saxátilé* (Heath Bed-straw).—*Leaves* about 6 in a whorl, inversely egg-shaped, pointed; *stem* much branched, smooth, prostrate below.—Heathy places, abundant. A small species, with numerous dense panicles of white flowers. The edges of the leaves are sometimes fringed with a few prickles, which point forwards. This and *G. verum* are the only British species which can be called ornamental.—Fl. June, August. Perennial.

* Besides the above, seven or eight other species belonging to this group have been described by British botanists; they are, however, of rare occurrence, with the exception of *G. tricorné*, which is tolerably common in some of the English counties, and in the Isle of Wight. The *stems* of this species are about a foot long, and rough, as well as the leaves, with prickles pointing backwards; the *flowers* grow in threes, and the *fruit* is reflexed.

†† *Fruit rough.*

7. *G. Aparíné* (Goose Grass, Cleavers).—*Leaves* 6—8 in a whorl, very rough with prickles pointing backwards; *flowers* 2—3 together, axillary; *fruit* rough with hooked prickles.—Hedges, exceedingly common. Well distinguished by its rough stems and leaves, which cling to the fingers when touched. The globular seed-vessels are also very tenacious, and disperse themselves by clinging to the coat of any animal that touches them; hence they derive their popular name of Cleavers. The whole plant, when green, is greedily devoured by geese.

x

The seeds, it is said, have been substituted for Coffee.—
Fl. June—August.   Annual.

8. *G. boreále* (Cross-leaved Bed-straw).—*Leaves* 4 in
a whorl, 3-nerved, smooth; *stem* erect; *fruit* rough, with
hooked prickles.—Damp rocky places in the North.
Well distinguished by its cruciform, smooth leaves, and
prickly fruit.—Fl. July, August.   Perennial.

ASPERULA ODORATA (*Sweet Woodruff*).

3. Aspérula (*Woodruff*).

1. *A. odoráta* (Sweet Woodruff).—*Leaves* about 8 in
a whorl; *flowers* in stalked, terminal panicles; *fruit*

rough with bristles.—Moist woods, common. A deservedly favourite plant, not so much for its regularly
whorled, bright green leaves, and pretty snow-white
flowers, as for its agreeable perfume when dry, which
resembles new hay. When it is desired to preserve the
leaves merely for their scent, the stem should be cut
through just above and below a joint, and the leaves
pressed in such a way as not to destroy their star-like
arrangement.—Fl. May, June. Perennial.

2. *A. Cynánchica* (Squinancy-wort).—*Leaves* 4 in a
whorl, linear, uppermost very unequal.—Dry pastures,
especially on a chalky or limestone soil, not common.
A small plant, with very narrow leaves, and tufts of
lilac or whitish flowers. It derives its name from
having been formerly used as a remedy for the squi-
nancy, or quinsy.—Fl. June, July. Perennial.

SHERARDIA ARVENSIS (*Field Madder*).

4. SHERARDIA (*Field Madder*).

1. *S. arvensis* (Field Madder).—A small plant, with
branched spreading *stems*, narrow pointed *leaves*, in
whorls of about 6 each, and minute bluish pink *flowers*,

which form a small umbel in the bosom of the terminal leaves.—Abundant in cultivated land.—Fl. June — August. Annual.

ORD. XLIV.—VALERIANEÆ.—THE VALERIAN TRIBE.

Calyx superior, finally becoming a border, or pappus, to the fruit; *corolla* tubular, 3—6-lobed, sometimes irregular and spurred at the base; *stamens* 1—5, inserted into the tube of the corolla; *ovary* with 1—3 cells; *fruit* dry, crowned with the calyx, not bursting, 1-seeded, two of the cells being empty.—Herbaceous plants, with opposite leaves, often strong-scented, or aromatic, inhabiting temperate countries, especially the north of India, Europe, and South America. Many of the plants in this order possess properties worthy of notice, but by far the most remarkable is *Valeriana Jatamansi*, the Spikenard of Scripture, and the *Nardus* of the ancient classical authors. It grows on the hills of Butan, in India, where it is called *Dshatamansi.* The root leaves, shooting up from the ground, and surrounding the young stem, are torn up, along with a part of the root, and having been dried in the sun, or by artificial heat, are sold as a drug. Two merchants of Butan, of whom Sir W. Jones caused inquiries to be made, related, that the plant shoots up straight from the earth, and that it is then, as to colour, like a green ear of wheat; that its fragrance is pleasant, even while it is green, but that its odorous quality is much strengthened by merely drying the plant; that it grows in Butan on hills, and even on plains, in many places; and that in that country it is gathered and prepared for medicinal purposes. In ancient times, this drug was conveyed by way of Arabia to southern Asia, and thus it reached the Hebrews. Judas valued the box of ointment with which Mary anointed our blessed Lord's feet at two hundred denarii (6*l.* 9*s.* 2*d.*) By the Romans, it was considered so precious, that the poet Horace promises to Virgil a whole *cadus*, or about three dozen modern bottles, for a small onyx-box full of spikenard. It was a Roman custom, in festive banquets, not only to crown the guests with flowers, but also to anoint them with spikenard. Eastern nations procure from the mountains of Austria the *Valeriana Celtica* and *V. Saliunca* to perfume their baths. Their roots are grubbed up with danger and difficulty by the peasants of Styria and Carinthia, from rocks on the borders of eternal snow; they are then tied in bundles, and sold at a very low price to merchants, who send them by way of Trieste to Turkey and Egypt, where they are retailed at a great profit, and passed onwards to the nations of India and Ethiopia. The seeds of *V. rubra* were used in former times in the process of embalming the dead; and some, thus employed in the 12th century, on being removed from the cere-cloth in the present century and planted, have vegetated. The roots of our common Valerian, *V. officinalis*, are still used in medicine; their effect on cats is very remarkable, producing a kind of intoxication. The young leaves of *Fedia olitoria* (Lamb's Lettuce) are eaten as salad, and those of *V. rubra* are, in Sicily, eaten in the same way.

1. VALERIANA (Valerian).—*Corolla* 5-cleft, bulged or spurred at the base; *stamens* 1—3; *fruit* crowned with a pappus. (Name from the Latin *valeo*, to be powerful, on account of its medicinal virtues.)

2. FÉDIA (Corn Salad).—*Corolla* 5-cleft, bulged at the base; *stamens* 3; *fruit* crowned with the calyx. (Name of uncertain origin.)

1. VALERIANA (*Valerian*).

1. *V. rubra* (Red Valerian).—*Corolla* spurred at the base; *stamen* 1; *leaves* egg-shaped, pointed.—Not a native plant, but nevertheless very common in limestone quarries and chalk-pits, as well as on old garden walls. The stems are 1—2 feet high; the leaves large, smooth,

and glaucous; the flowers are deep red, and grow in terminal bunches. A white variety is not uncommon.—Fl. June—September. Perennial.

VALERIANA RUBRA (*Red Valerian*).

2. *V. dioíca* (Small Marsh Valerian).—*Stamens* and *pistils* on different plants; *corolla* bulged at the base; *stamens* 3; *root leaves* egg-shaped, stalked; *stem leaves* pinnatifid, with a large terminal lobe.—Marshy ground, not unfrequent. Growing about a foot high, quite erect,

and unbranched, with a terminal corymb of light pink flowers. The flowers which bear stamens only are the largest.—Fl. May. Perennial.

3. *V. officinalis* (Great Wild Valerian).—*Corolla* bulged at the base; *stamens* 3; *leaves* pinnate.—Much taller and stouter than the last, but resembling it in habit, as well as in colour and smell of the flowers. This is the species of which the roots are used in medicine, and of which cats are so fond; rats also are said to be attracted by their smell.—Fl. June, July. Perennial.

FEDIA OLITORIA (*Common Corn Salad, Lamb's Lettuce*).

2. FÉDIA (*Corn Salad.*)

1. *F. olitória* (Common Corn Salad, Lamb's Lettuce). —*Leaves* long and narrow, wider towards the end, a

little toothed near the base; *flowers* in leafy heads; *capsule* inflated, crowned by the 3 calyx teeth.—Cultivated ground, abundant. A small plant 4—8 inches high, with tender bright green leaves, stems repeatedly 2-forked, and terminal leafy heads of very minute flowers, which resemble white glass. It is sometimes cultivated as a salad.—Fl. May, June. Annual.

2. *F. dentata* (Toothed Corn-salad).—*Leaves* long and narrow, much toothed towards the base; *flowers* in corymbs, with a solitary sessile one in the forks of the stem; *capsule* not inflated, crowned by the 4-toothed calyx.—Cultivated ground, common in the West of England, but not general elsewhere. Taller than the last, and of a more rigid habit.—Fl. June, July. Annual.

* Two or three other species of *Fedia* have been described by British botanists; these are of rare occurrence, and differ slightly from the preceding, principally in the form of the fruit.

ORD. XLV.—DIPSACEÆ.—THE TEAZEL TRIBE.

Calyx attached to the ovary, surrounded by several, more or less rigid, calyx-like bracts; *corolla* tubular, with 4—5 unequal lobes; *stamens* 4, the anthers not united; *style* 1; *stigma* not cleft; *fruit* dry, 1-seeded, crowned by the pappus-like calyx; *flowers* in heads.— Herbaceous plants, inhabiting temperate regions, and possessing no remarkable properties. *Dipsacus Fulló;um* is the Clothiers' Teazel, a plant with large heads of flowers, which are imbedded in stiff, hooked bracts. These heads are set in frames and used in the dressing of broad-cloth, the hooks catching up and removing all loose particles of wool, but giving way when held fast by the substance of the cloth. This is almost the only process in the manufacture of cloth which it has been found impossible to execute by machinery, for although various substitutes have been proposed, none has proved, on trial, exactly to answer the purpose intended.

1. Dípsacus (Teazel).—*Heads* with numerous general bracts at the base ; *outer calyx* forming a thickened margin to the fruit ; *inner*, cup-shaped, entire ; *receptacle* bristling with rigid awns ; *fruit* with four sides. (Name ; the leaves are united at their base, so as to form, round the stem, a hollow, in which water collects ; hence the plant was called *dípsacus*, or thirsty, from the Greek *dipsao*, to thirst.)

2. Scabiosa (Scabious). — *Heads* with numerous general bracts at the base ; *outer calyx* membranaceous and plaited ; *inner* of 5 bristles ; *receptacle* scaly ; *fruit* nearly cylindrical. (Name, from the Latin *scabies*, the leprosy, for which disease some of the species were supposed to be a remedy.)

3. Knautia (Field Scabious).—*Heads* with numerous general bracts at the base ; *outer calyx* minute, with 4 small teeth ; *inner* cup-shaped ; *receptacle* hairy ; *fruit* 4-sided. (Named in honour of Christopher Knaut, a Saxon botanist.)

1. Dípsacus (*Teazel*).

1. *D. sylvestris* (Wild Teazel).—*Leaves* opposite, united at the base, and forming a cup ; *bristles* of the receptacle not hooked.—Waste places, common. A stout herbaceous plant, 3—6 feet high, with an erect prickly stem, large bright green leaves, which are prickly underneath and united at the base, and often contain water. The flowers grow in large, conical, bristly heads, the terminal bristles being generally the longest. The flowers themselves are light purple, and expand in irregular patches on the head.—Fl. July. Biennial.

2. *D. pilósus* (Small Teazel).—*Leaves* stalked, with a small leaflet at the base on each side.—Moist shady places, not common. Smaller than the last in all its parts, and having more the habit of a Scabious than of

a Teazel.   The flowers are white, and grow in small,
nearly globose, bristly heads ; the whole plant is rough
with bristles.—Fl. August, September.   Biennial.

    * *D. Fullónum* (Fullers' Teazel) differs from *D. syl-*
vestris in having the *bristles* of the *receptacle* hooked : it
is not considered a British plant, though occasionally
found wild in the neighbourhood of the cloth districts.

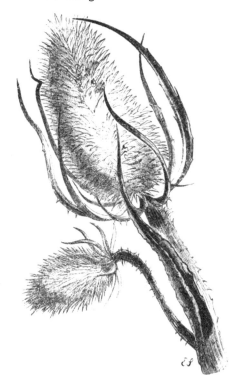

DIPSACUS SYLVESTRIS (*Wild Teazel*).

SCABIOSA SUCCISA (*Premorse Scabious*).

2. SCABIOSA (*Scabious*).

1. *S. succisa* (Premorse Scabious).—*Corolla* 4-cleft, nearly regular; *heads* nearly globose; *leaves* oblong, entire.—Heaths and pastures, common. A slender, little-branched plant, with a hairy stem, few leaves, and terminal heads of purplish blue flowers. The root is solid and abrupt, as if bitten off (premorse).—Fl. July—October. Perennial.

2. *S. columbária* (Small Scabious).—*Corolla* 5-cleft, the outer flowers longest; *heads* nearly globose; *root leaves* oblong, variously cut, *upper* pinnatifid.—Pastures on a chalky soil, not uncommon. Well distinguished from the last by its radiate flowers and cut leaves. The foliage is of a much lighter hue, and the flowers lilac rather than purple.—Fl. July, August. Perennial.

KNAUTIA ARVENSIS (*Field Scabious*).

3. KNAUTIA (*Field Scabious.*)

1. *K. arvensis* (Field Scabious).—*Root leaves* simple; *upper* pinnatifid; *inner calyx* with a fringe of 8—16

teeth ; *heads* convex.—Corn-fields and way-sides, common. A tall bristly plant, 2—3 feet high, not much branched, bearing several large handsome heads of lilac flowers, of which the inner are regularly cleft, the outer larger, and generally, though not always, radiate. It is sometimes called *Scabiosa arvensis.*—Fl. July, August. Perennial.

Ord. XLVI.—COMPÓSITÆ.—Compound Flowers.

This extensive and well-marked Order derives its name from having its flowers *compounded,* as it were, of numerous smaller ones called *florets,* which are inclosed within a calyx-like assemblage of *bracts,* termed an *involucre.* These bracts, usually called *scales,* often overlap one another like the tiles of a house (*imbrex*); hence they are said to be *imbricated.* The flowers vary greatly in shape, but the following description will be found to include all the British species. *Calyx* rising from the top of the ovary, and becoming a *pappus,* that is, either a chaffy margin of the fruit, or a tuft, or ring of bristles, hairs, or feathery down ; *corolla* of one petal, either tubular, or strap-shaped ; *stamens 5,* united by their anthers (*syngenesious*) ; *ovary* inferior, 1 to each style, 1-celled ; *style* simple, with a simple or 2-cleft *stigma,* sheathed by the tube of anthers ; *fruit* a solitary erect seed, crowned by the *pappus,* which is sometimes merely a chaffy margin, but more frequently an assemblage of simple, or serrated, (feathery) hairs, sometimes elevated on a stalk. For convenience of reference, this Order is divided into several groups ; I. CICHORÁCEÆ (*Chicory Group*). In this, all the florets are strap-shaped and perfect ; that is, each contains 5 stamens and a pistil. The prevailing colour of British species is yellow, as the Dandelion ; but Salsafy (*Tragopógon porrifolius*) and Alpine Sow-Thistle (*Sonchus alpinus*) have purple flowers ; Chicory, blue. In II. CYNAROCÉPHALÆ (*Thistle Group*) the florets form a *convex* head, and are all tubular and perfect, except in *Centauréa*, where the outer florets, which are larger than the inner, are destitute both of stamens and pistils ; the stigma is jointed on the style. The flowers are purple, with a tendency to vary into white ; but in Carline-Thistle *Carlina*) they are brownish yellow ; in Corn-flower (*Centauréa Cýanus*) bright blue. In III. TUBIFLORÆ (*Tansy Group*) all the *florets* are tubular and perfect, and form a *flat* head ; the *style* passes into the stigma without a joint ; the *flowers* are mostly yellow ; but Hemp-Agrimony (*Eupatorium cannábinum*) has lilac flowers ; Butter-bur (*Petasítes vulgáris*) pale flesh-coloured, and in most species of *Artemisia, Gnaphalium,* and *Filágo,* the colour is determined rather by the involucre than the florets. In IV. RADIATÆ (*Daisy Group*) the *florets* are of two kinds ; those of the centre, or *disk,* being tubular and perfect, those of the margin, or *ray,* strap-shaped, and having pistils only. The prevailing colour of the disk is yellow, Yarrow (*Achilléa*) being the only exception, in which all the florets are white ; the ray is either of the same colour, as in Colt's-foot (*Tussilago*), Golden-rod (*Solidago*), Rag-wort (*Senecio*), Flea-wort (*Cineraria*), Leopard's-Bane (*Doronícum*), Elecampane (*Ínula*), Flea-bane (*Pulicaria*), Corn-Marigold (*Chrysánthemum ségetum*), and Ox-eye Cha-momile (*Ánthemis tinctoria*) ; white, as in Daisy (*Bellis*), Feverfew (*Pýrethrum*), Ox-eye (*Chrysánthemum Leucán-themum*), *Matricaria,* and several species of Chamomile (*Ánthemis*) ; or purple, as in Starwort (*Aster*), and *Erigeron.* In Groundsel (*Senecio vulgáris*) the ray is never perfected.

The limits of the Order COMPÓSITÆ are exactly the same as those of the Linnean Class SYNGENESIA ; but the number of plants belonging to it exceeds the amount of *all the plants known to Linnæus,* so extensive have been the researches in Botany since his time. According-ing to Lindley, the number of *genera* alone amounts to 1,005 ; of *species* to 9,000 ; the whole number of plants

known to Linnæus being but 8,500. It is supposed
that the compound flowers constitute about one-tenth of
all described plants; their proportion varies considerably in different parts of the world, but not according
to any fixed rule. The Chicory Group are, however, most
abundant in cold regions, the Daisy Group in hot
climates. Again, it may be remarked, that, in cold and
temperate regions, the Compósitæ are mostly herbaceous,
but as we approach the equator they become shrubs, or
even trees. The variety of properties which they possess is not proportionate to the immense number of
species. Bitterness, in a greater or less degree, is a
characteristic of nearly all, to which is sometimes added
astringency, and many possess tonic or narcotic properties. Chicory, or Succory, is cultivated as a salad, but
more frequently for the sake of its roots, which are
roasted and mixed with ground coffee. The flavour is
agreeable, but it is to be feared that less palatable and,
perhaps, not so wholesome roots, procurable at a less
cost, are often substituted for it. From the leaves a
blue dye may be obtained. Endive is another species
of Chicory (*Cichorium Endivia*), the blanched leaves of
which afford a common winter salad. The common Dandelion (a corruption of *Dent-de-lion*, " Lion's tooth,") supplies an extract which is said to have valuable medicinal
properties ; its roots are also used to adulterate coffee.
Lettuces afford a wholesome salad, as well as an extract,
the properties of which resemble those of opium. The
roots of *Scorzonéra* and *Tragopógon porrifolius* (Car-
doons) are esculent, but almost grown out of use.
These all belong to the Chicory Group.

Among the Thistle Group we have the Artichoke
(*Cýnara Scólymus*), the young involucres and recep-
tacles of which are edible ; the Burdock (*Lappa*), the
root of which is said to be useful in rheumatism ; and
the Carline Thistle, which was anciently used in magi-
cal incantations. In the third Group, Wormwood
(*Artemisia*) is remarkable for its intense bitterness.

One species (*A. Abrótanum*) is the Southernwood of gardens, a fragrant shrub, used on the Continent in making beer ; (*A. Dracúnculus*), the Tarragon of gardeners, is used for giving an agreeable flavour to vinegar. Some species of *Eupatorium* have the reputed power of healing the bites of venomous animals ; and *E. glutinosum* is said to be the plant which, under the name of Mático, is extensively used as a styptic. It is a shrubby plant, inhabiting the Andes, and derived its name from·a soldier, nicknamed " Matico" (little Matthew), who having been wounded in battle, accidentally applied the leaves of this plant to his wound, which had the immediate effect of stopping the bleeding. To the RADIATÆ belong the gorgeous Dahlia, so called, from Dr. Dahl, who introduced it ; and the "wee" Daisy, or Day's-eye, which opens only in sunny weather, and peeps up through the grass as if it were an eye indeed. The genus Helianthus contains the Sun-flower (*H. annuus*), and Jerusalem Artichoke (*H. tuberosus*) ; " Jerusalem" being a corruption of an Italian word (*girasole*), of the same meaning as Sun-flower, the name Artichoke being given to mark the similarity of flavour in its roots with that of the true Artichoke mentioned above. It never flowers in England, but produces abundance of tubers, which hold a high rank among esculent vegetables. It is valuable, not only for its productiveness, but for the freedom with which it grows in any soil. Its roots are sometimes made into a dish, which, by an absurd piece of pedantry, is called " Palestine soup." Chamomile and Feverfew possess valuable medicinal properties, especially the former. Coltsfoot and Elecampane are useful in pectoral complaints ; the flowers of Marigold are used to adulterate saffron ; the Ox-eye Daisy is said to be destructive to fleas ; the yellow Ox-eye affords a yellow dye, and the petals of the Dahlia a beautiful carmine.

I. CICHORACEÆ.—*Chicory Group*.

All the florets strap-shaped, having stamens and pistils.

II. CYNAROCÉPHALÆ.—*Thistle Group*.

All the florets tubular, 5-cleft, having stamens and pistils, (except in CENTAURÉA, in which the outer florets are larger, and destitute of stamens and pistils,) *and forming a convex head ; style jointed below the stigma.*

III. TUBIFLORÆ.—*Tansy Group*.

All the florets tubular, 5-cleft, having stamens and pistils, and forming a flat head ; style not jointed below the stigma.

IV. RADIATÆ.—*Daisy Group*.

Central florets tubular, 5-cleft, having stamens and pistils ; outer florets strap-shaped, forming a ray, and furnished with pistils only. (*Senecio vulgaris*, Common Groundsel, has no rays.)

I. CICHORACEÆ.—*Chicory Group*.

1. TRAGOPÓGON (Goat's-beard).—*Involucre* simple, of 8—10 long scales, united below ; *receptacle* dotted ; *fruit* rough, with longitudinal ridges, tapering into a long beak ; *pappus* feathery, with the down interwoven, concave above. (Name in Greek signifying *a goat's beard*.)

2. HELMINTHIA (Ox-tongue). — *Involucre* of about 8 equal scales, surrounded by 3—5 leaf-like, loose bracts ; *receptacle* dotted ; *fruit* rough, with transverse wrinkles, rounded at the end and beaked ; *pappus* feathery. (Name from the Greek *helmins, helminthos*, a worm, from the form of the fruit.)

Y

3. PICRIS.—*Involucre* of 1 row of equal, upright scales, with several small spreading ones at the base ; *receptacle* lightly dotted ; *fruit* rough, with transverse ridges, not beaked ; *pappus* of 2 rows, the inner only feathery. (Name from the Greek *picros*, bitter.)

4. APARGIA (Hawk-bit).—*Involucre* unequally imbricated, with the outer scales smaller, black, and hairy, in several rows ; *receptacle* lightly dotted ; *fruit* tapering to a point ; *pappus* of 1 row, feathery.   (Name of uncertain origin.)

5. THRINCIA.—*Involucre* of 1 row, with a few scales at the base ; *receptacle* lightly dotted ; *fruit* of the outer florets scarcely beaked ; *pappus* a chaffy fringed crown ; *fruit* of the inner florets beaked ; pappus feathery. (Name from the Greek *thrincos*, a battlement, from the form of the seed-crown of the marginal florets.)

6. HYPOCHÆRIS (Cat's-ear).—*Involucre* oblong, imbricated ; *receptacle* chaffy ; *fruit* rough, often beaked ; *pappus* feathery, often with a row of short bristles outside.   (Name in Greek denoting its fitness for hogs.)

7. LACTÚCA (Lettuce).—*Involucre* oblong, imbricated, its scales membranous at the margin, containing but few flowers ; *receptacle* naked ; *fruit* flattened, beaked ; *pappus* hairy. (Name from *lac*, milk, which the juice resembles in colour.)

8. SONCHUS (Sow-thistle). — *Involucre* imbricated, with 2 or 3 rows of unequal scales, swollen at the base ; *receptacle* naked ; *fruit* flattened, transversely wrinkled, not beaked ; *pappus* hairy. (Name, in Greek, bearing allusion to the soft nature of the stems.)

9. CREPIS (Hawk's-beard).—*Involucre* double, inner of 1 row, outer of short, loose scales ; *receptacle* naked ; *fruit* not flattened, furrowed, tapering upwards ; *pappus*, a tuft of soft, white down.   (Name in Greek signifying a *slipper*, but why given to this plant is not known.)

10. HIERÁCIUM (Hawk-weed).—*Involucre* imbricated with numerous oblong scales ; *receptacle* dotted ; *fruit*

angular, furrowed, abrupt, with a toothed margin at the top ; *pappus* bristly, sessile, not white. (Name from the Greek *hierax*, a hawk, because that bird was supposed to use the plant to strengthen its sight.)

11. LEÓNTODON (Dandelion).—*Involucre* imbricated with numerous scales, the outermost of which are loose, and often reflexed ; *receptacle* dotted ; *fruit* slightly flattened, rough, bearing a long and very slender beak ; *pappus* hairy. (Name from the Greek, *leon*, a lion, and *odous, odontos*, a tooth, from the tooth-like lobes of the leaves.)

12. LAPSÁNA (Nipple-wort).—*Involucre* a single row of erect scales, with 4—5 small ones at the base, containing but few flowers ; *receptacle* naked ; *fruit* flattened, furrowed ; *pappus* 0. (Name of Greek origin.)

13. CICHORIUM (Chicory).—*Involucre* in 2 rows, inner of 8 scales, which bend back after flowering ; outer of 5 smaller loose scales ; *receptacle* naked, or slightly hairy ; *fruit* thick above, tapering downwards ; *pappus* a double row of small chaffy scales. (Name of Arabic origin.)

II.—CYNAROCÉPHALÆ.—*Thistle Group.*

14. ARCTIUM (Burdock).—*Involucre* globose, scales ending in hooked points ; *receptacle* chaffy ; *fruit* oblong, 4-sided ; *pappus* short. (Name from the Greek *arctos*, a bear, from the roughness of the heads of flowers.)

15. SERRÁTULA (Saw-wort).—*Stamens* and *pistils* on different plants ; *involucre* imbricated, scales not prickly ; *receptacle* chaffy or bristly ; *fruit* flattened, not beaked ; *pappus* hairy. (Name from the Latin *sérrula*, a little saw, the leaves being finely serrated.)

16. SAUSSÚREA.—*Involucre* imbricated, scales not prickly ; *anthers* bristly at the base ; *receptacle* chaffy ; *pappus* double, outer bristly, inner longer, feathery. (Named in honour of the two *Saussures*, eminent botanists.)

17. CARDUUS (Thistle).—*Involucre* swollen below, imbricated with thorn-like scales ; *receptacle* bristly ; *pappus* hairy, united by a ring at the base, and soon falling off. (The Latin name of the plant.)

18. CNICUS (Plume-Thistle).—Resembling *Carduus*, except that the *pappus* is feathery. (Name from the Greek *cnizo*, to prick.)

19. ONOPORDIUM (Cotton-Thistle).—*Receptacle* honey-combed ; *fruit* 4-angled ; *pappus* hairy, rough ; in other respects resembling *Carduus*. (Name of Greek origin.)

20. CARLÍNA (Carline-Thistle).—Resembling *Cnicus*, except that the inner scales of the *involucre* are chaffy and coloured, and spread like a ray. (Name, the same as *Carolina*, from a tradition that the root of one species, *C. acaulis*, was shown by an angel to *Charlemagne* as a remedy for the plague, which prevailed in his army.)

21. CENTAURÉA (Knapweed, Blue-bottle, &c.)—*Involucre* imbricated ; *receptacle* bristly ; *pappus* hairy, or 0 ; *outer florets* large, irregular, destitute of stamens and pistils. (Name from the *Centaur Chiron*, who is fabled to have healed wounds with it.)

III. TUBIFLORÆ.—*Tansy Group*.

22. BIDENS (Bur-Marigold).—*Fruit* covered with 2 or 3 erect rigid bristles, which are rough with minute teeth, pointing downwards. (Name from the Latin *bis*, double, and *dens*, a tooth, from the structure of the fruit.)

23. EÚPATORIUM (Hemp-Agrimony). — *Heads* few flowered ; *involucre* imbricated, oblong ; *receptacle* naked ; *styles* much longer than the florets. (Name from *Mithridátes Eúpator*, who is said to have brought the plant into use.)

24. CHRYSÓCOMA (Goldylocks).—*Involucre* a single row of loosely spreading scales ; *receptacle* honey-combed ;

fruit flattened, silky; *pappus* hairy, rough. (Name from the Greek *chrysos*, gold, and *comé*, hair.)

25. Diótis (Cotton-weed).—*Pappus* 0 ; *corolla* with two ears at the base, which remain and crown the fruit. (Name from the Greek *dis*, double, and *ous*, *ótos*, an ear, from the structure of the fruit.)

26. Tanacétum (Tansy.)—*Involucre* cup-shaped, imbricated ; *receptacle* naked ; *fruit* crowned with a chaffy border. (Name altered from the Greek *athánaton*, everlasting.)

27. Artemisia (Wormwood).—*Pappus* 0 ; *involucre* roundish, imbricated, containing but few flowers. (Named after *Ártemis*, the Diana of the Greeks.)

28. Gnaphalium (Cudweed). —*Involucre* roundish, dry, imbricated, often coloured ; *receptacle* naked ; *pappus* hairy ; *stamens* and *pistils* sometimes on different plants. (Name from the Greek *gnaphálion*, soft down, with which the leaves are covered.)

29. Filágo.—*Involucre* tapering upwards, imbricated, of a few long pointed scales ; *receptacle* chaffy in the circumference ; *pappus* hairy ; *florets* few, the outer ones bearing pistils only. (Name from the Latin *filum*, a thread, the whole plant being clothed with white thread-like hairs.)

30. Petasítes (Butter-Bur).—*Involucre* a single row of narrow scales ; *receptacle* naked ; *stamens* and *pistils*, for the most part, on different plants. (Name from the Greek *pétasos*, a covering for the head, from the large size of the leaves.)

IV. Radiatæ.—*Daisy Group.*

31. Tussilágo (Colt's-foot).—*Involucre* a single row of narrow scales ; *receptacle* naked ; *florets* of the ray narrow, in several rows ; of the disk few, all yellow. (Name from the Latin *tussis*, a cough, from the use to which it is applied.)

32. Erígeron (Flea-bane).—*Involucre* imbricated

with narrow scales; *receptacle* naked; *florets* of the ray
in many rows, very narrow, different in colour from
those of the disk. (Name in Greek signifying *growing
old at an early season,* from the early appearance of the
grey seed-down.)

33. ASTER (Starwort).—*Involucre* imbricated, a few
scales on the flower-stalk; *receptacle* naked, honey-
combed; *florets* of the ray in 1 row, purple; of the disk
yellow; *pappus* hairy, in many rows. (Name from the
Greek *aster,* a star.)

34. SOLIDÁGO (Golden-rod).—*Involucre* and *receptacle*
as in ASTER; *florets* all yellow; *pappus* hairy, in 1 row.
(Name from the Latin *solidare,* to unite, on account of
its supposed property of healing wounds.)

35. SENÉCIO (Ragwort and Groundsel).—*Involucre*
imbricated, oblong, the scales tipped with brown, a few
smaller ones at the base; *receptacle* naked; *florets* all
yellow, the outer in *S. vulgaris* wanting. (Name from
the Latin *senex,* an old man, from the grey seed-down.)

36. CINERARIA (Flea-wort).—*Involucre* oblong, of
1 row of equal scales; *receptacle* naked; *florets* all
yellow. (Name from the Latin *cineres,* ashes, from the
pale hue of the under side of the leaves in some species.)

37. DORÓNICUM (Leopard's-bane).—*Involucre* cup-
shaped, scales equal, in 2 rows; *florets* all yellow;
pappus hairy, wanting in the florets of the ray. (Name
of uncertain etymology.)

38. ÍNULA (Elecampane).—*Involucre* imbricated, in
many rows; *receptacle* naked; *florets* all yellow; *anthers*
with two bristles at the base; *pappus* hairy, in 1 row.
(Name probably a corruption of *Helénula,* Little Helen.)

39. PULICARIA (Flea-bane).—*Involucre* loosely imbricated, in few rows; *pappus* in 2 rows, outer one short,
cup-shaped, toothed, inner hairy, in other respects like
ÍNULA. (Name from the Latin *pulex,* a flea, which is
supposed to be driven away by its powerful smell.)

40. BELLIS (Daisy).—*Involucre* of 2 rows of equal
blunt scales; *receptacle* conical; outer *florets* white,

inner yellow ; *pappus* 0. (Name from the Latin *bellus*, pretty.)

41. CHRYSÁNTHEMUM (Ox-eye).— *Involucre* nearly flat, the scales membranaceous at the margin ; *receptacle* naked ; *pappus* 0. (Name from the Greek *chrysos*, gold, and *anthos*, a flower.)

42. PÝRETHRUM (Feverfew).—*Involucre* cup-shaped, the scales membranaceous at the margin ; *receptacle* naked ; *florets* of the ray white, of the disk yellow; *pappus* a membranaceous border. (Name from the Greek *pýrethron*, the name of some plant.)

43. MATRICARIA (Wild Chamomile).—*Involucre* nearly flat, the scales not membranaceous at the margin ; *receptacle* conical, naked; *florets* of the ray white, of the disk yellow ; *pappus* 0. (Name from some supposed medicinal virtues.)

44. ÁNTHEMIS (Chamomile).—*Involucre* cup-shaped, or nearly flat, the scales membranaceous at the margin ; *receptacle* convex, chaffy ; *pappus* 0, or a narrow chaffy border. (Name from the Greek *anthos*, a flower, from the value of its blossoms as a medicine.)

45. ACHILLÉA (Yarrow).—*Involucre* egg-shaped or oblong, imbricated ; *receptacle* flat, chaffy ; *florets* all of one colour, those of the ray 5—10, broad ; *pappus* 0. (Named after Achilles.)

I. CICHORACEÆ.—*Chicory Group*.

1. TRAGOPÓGON (*Goat's-beard*).

1. *T. pratensis* (Yellow Goat's-beard).—*Involucre* about as long as, or longer than, the corolla ; *leaves* broad at the base, very long, tapering, channelled, undivided ; flower-stalks slightly thickened above.— Meadows, not uncommon. An erect glaucous plant, about 2 feet high, with long grass-like leaves, and large bright yellow flowers, which always close early in the day, and have hence gained for the plant the name of John-go-to-bed-at-noon. The pappus is very beautiful,

the feathery down being raised on a long stalk, and interlaced so as to form a kind of shallow cup.—Fl. June, July. Biennial.

TRAGOPOGON PRATENSIS (*Yellow Goat's-beard*).

     * *T. porrifólius* (Salsafy).—Though not a British species, is occasionally found in moist meadows. In habit it resembles the last, but has purple flowers. It

was formerly much cultivated for the sake of its fleshy, tapering roots, which were boiled or stewed, and eaten. Its place is now supplied by *Scorzonéra Hispánica :* but directions for its culture are still given in most gardeners' calendars.

HELMINTHIA ECHIOIDES (*Bristly Ox-tongue*).

2. HELMINTHIA (*Ox-tongue*).

1. *H. echioídes* (Bristly Ox-tongue).—Waste places, not uncommon. A stout and much branched herb, 2—3 feet high, well distinguished by its numerous *prickles*, each of which springs from a raised white spot, and by the large heart-shaped *bracts* at the base of the yellow *flowers.*—Fl. June, July. Perennial.

3. PICRIS.

1. *P. hieracioídes* (Hawk-weed Picris).—Waste places, common. A rather slender plant, 2—3 feet high, branched principally above ; the *stems* are rough with hooked bristles ; the *leaves* narrow, rough, and toothed,

sessile ; the *flowers* numerous, yellow, with several scales on the stalks.—Fl. July—September.　Biennial.

PICRIS HIERACIOIDES (*Hawk-weed Picris*).

4. APARGIA (*Hawk-bit*).

1. *A. hispida* (Rough Hawk-bit).—*Leaves* all from the root, pinnatifid, with the lobes pointing backwards, rough with forked bristles ; *stalk* single-flowered.— Meadows and waste places, frequent.—*Flowers* yellow.— Fl. June—September.　Perennial.

2. *A. autumnalis* (Autumnal Hawk-bit).—*Leaves* all from the root, narrow, slightly hairy on the ribs beneath ; *stalk* many flowered, swollen beneath the flowers.—Meadows and cornfields, frequent.　A tall plant, 2—3 feet

high, with a downy *involucre,* and large deep-yellow *flowers.*—Fl. August. Perennial.

APARGIA AUTUMNALIS (*Autumnal Hawk-bit*).

5. THRINCIA.

1. *T. hirta* (Hairy Thrincia).—Heaths and downs, common. A small plant, 4—6 inches high, with spreading, more or less lobed, *leaves,* which are rough with

forked bristles, and leafless somewhat hairy *stalks*, each of which bears a yellow *flower.*—Fl. July—Sept. Perennial.

THRINCIA HIRTA (*Hairy Thrincia*).

6. HYPOCHÆRIS (*Cat's-ear*).

1. *H. radicata* (Long-rooted Cat's-ear).—Leaves all from the root, pinnatifid, with the lobes pointing backwards, bristly; *stalks* branched, smooth, with a few scales below the flowers.—Hedges and waste places, common. Well distinguished by its long branched flower-stalks, which are quite smooth throughout, and slightly swollen beneath the large yellow flowers, where there are also a few small scales.—Fl. July, August. Perennial.

HYPOCHÆRIS RÁDICATA (*Long-rooted Cat's-ear*).

* Besides the above, there are two other British species of Hypochæris : *H. glabra* (Smooth Cat's-ear), which has nearly smooth *leaves*, and small yellow *flowers* scarcely larger than the *involucre ;* gravelly places, not common; and *H. maculata* (spotted Cat's ear), which has rough spreading leaves, which are not lobed, and 1 (or rarely more) very large deep yellow *flowers ;* limestone and magnesian rocks, rare.

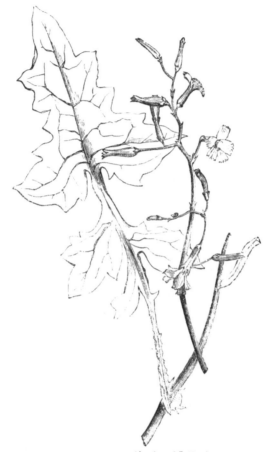

LACTUCA MURALIS (*Ivy-leaved Lettuce*).

7. LACTÚCA (*Lettuce*).

1. *L. murális* (Ivy-leaved Lettuce).—*Florets 5 ; leaves*
pinnatifid, variously cut, with the terminal lobe largest ;

panicles spreading.—Woods and old walls, not uncommon. A slender plant, leafy below, 1—2 feet high, with small yellow heads, each of which contains 5 regular florets, and thus resembles a simple flower of 5 petals. The panicle has a singularly angular growth ; the fruit is black.—Fl. July—September.   Biennial.

* There are three other British species of *Lactúca,* which are less common than the preceding : *L. Scaríola* (Prickly Lettuce), which has nearly perpendicular *leaves,* prickly at the back ; *L. virósa* (Strong-scented Lettuce), has spreading prickly *leaves ;* and *L. saligna* has narrow toothed *leaves,* and bears its *flowers* in lateral tufted clusters.   The Garden Lettuce (*L. satíva*) belongs to this genus, but is not a native plant.

8. SONCHUS (*Sow-thistle*).

1. *S. oleráceus* (Common Sow-thistle, Milk-thistle).— *Leaves* oblong, more or less pinnatifid or entire, toothed, often prickly, the upper ones clasping the stem ; *stem* branched ; *heads* somewhat umbellate ; *involucres* smooth.—Waste places, and as a weed in gardens, common.   This plant is commonly known under the name of Milk-thistle, and is much sought after as food for rabbits.—Fl. June—September.   Annual.

2. *S. arvensis* (Corn Sow-thistle, Milk-thistle).—*Leaves* oblong, the lower ones pinnatifid, with the lobes pointing backwards, toothed, heart-shaped at the base ; *stem* simple ; *heads* corymbose ; *involucre* and *flower-stalks* hairy.—In similar situations with the last, from which it may be readily distinguished by its simple stem, and much larger flowers.—Fl. August, September.   Perennial.

* *S. palustris* (Marsh Sow-thistle) is a much taller species than either of the preceding, growing 6—8 feet high ; the *leaves* are arrow-shaped at the base ; the *flowers* large and yellow, with hairy *involucres.*   It grows in marshy places ; rare.   *S. alpínus* (Alpine Sow-thistle)

is a very handsome plant, with large purple *flowers*. It grows on the Clova mountains but is rare.

SONCHUS OLERACEUS (*Common Sow-thistle*, *Milk-thistle*).

9. CREPIS (*Hawk's-beard*).

1. *C. virens* (Smooth Hawk's-beard).—*Leaves* smooth, pinnatifid, with the lobes pointing backwards, the upper ones narrow, arrow-shaped at the base, and clasping the stem ; *flowers* numerous.—In waste ground, and on the

roofs of cottages; common. Varying in height, from 6 inches to 2 feet, or more, and producing abundance of yellow flowers.—Fl. July—September. Annual.

CREPIS VIRENS (*Smooth Hawk's-beard*).

C. paludósa (Marsh Hawk's-beard).—*Leaves* smooth, the lower ones pinnatifid, with the lobes pointing backwards, tapering into a stalk; the upper ones narrow, heart-shaped at the base, and clasping the stem.—Damp woods; not unfrequent.—Fl. July—September. Perennial.

* There are several other British species of *Crepis*,

z

some of which are of common occurrence. For a description of these the student is referred to Hooker's British Flora.

10. Hieracium (*Hawk-weed*).

1. *H. Pilosella* (Mouse-ear Hawk-weed).—*Stem* single-flowered, leafless ; *scions* creeping ; *leaves* entire, hairy. —Banks, and dry pastures ; common. Well distinguished from all other British plants of the Order by its creeping scions, by its hairy undivided leaves, which are hoary underneath, and by its bright lemon-coloured flowers.—Fl. May—July. Perennial.

2. *H. murórum* (Wall Hawk-weed).—*Stem* many-flowered, with 1 leaf, branched above ; *root-leaves* egg-

shaped, toothed at the base.—Walls and rocks ; common. A very variable plant, 12—18 inches high, rarely having more than one leaf on the stem, sometimes none ; the leaves are usually more or less hairy, the flowers small, yellow.—Fl. July, August. Perennial.

3. *H. sylváticum* (Wood Hawk-weed).—*Stem* many-flowered, with a few leaves ; *leaves* narrow, egg-shaped, toothed, with the teeth pointing upwards ; *involucre* hoary with down.—Woods and banks ; common. A very variable plant, both in size and habit. The leaves are sometimes very slightly toothed, at other times deeply so, and often spotted with purple ; the flowers are large and yellow.—Fl. August, September. Perennial.

4. *H. Sabaudum* (Shrubby Hawk-weed).—*Stem* rigid, many-flowered, leafy ; lower *leaves* tapering into a short stalk, upper sessile, rounded at the base.—Woods and banks ; not unfrequent. As variable a plant as the last. —Fl. August, September. Perennial.

5. *H. umbellatum* (Narrow-leaved Hawk-weed).— *Stem* rigid, many-flowered, leafy ; *leaves* narrow, slightly toothed ; *flowers* in a terminal corymb ; *scales* of the *involucre* reflexed at the point.—Woods ; not unfrequent. A tall plant, 2—3 feet high, with a remarkably erect growth, unbranched, and terminating in an almost umbellate tuft of large, yellow flowers.—Fl. August, September. Perennial.

* Eight other species of Hawk-weed besides the above are described by Sir W. J. Hooker, some of which are doubtful natives, the others rare. Mr. Babington raises the number of British species to twenty-one, treating many as *species* which other botanists consider to be mere *varieties*. Of these it is not thought necessary to give even the names here, but most of them will agree with one or other of the descriptions given above. The student, however, must not be disheartened if he find himself occasionally at fault, the family being one of the most perplexing in this difficult Order.

LEONTODON TARAXACUM (*Common Dandelion*).

11. Leóntodon (*Dandelion*).

1. *L. Taráxacum* (Common Dandelion).—Cultivated ground ; abundant. Dandelion, from the French, *Dent-de-lion*, lion's-tooth, is the popular name of nearly all the larger yellow flowers belonging to this Sub-order. The true Dandelion may, however, be readily known by the following characters. The *leaves* all spring from the root, and are deeply cut, with the sharp lobes pointing backwards ; the *flower-stalks* are hollow, smooth, and leafless, and bear a single flower ; the *outer* scales of the *involucre* are reflexed ; the *pappus* is stalked, and white ; the *heads* when in fruit are of a globular form ; and the *receptacle*, after the fruit has been blown away, is convex, and dotted.—Fl. nearly all the year round. Perennial.

LAPSANA COMMUNIS (*Common Nipple-wort*).

12. LAPSÁNA (*Nipple-wort*).

1. *L. commúnis* (Common Nipple-wort). — *Leaves* stalked, toothed, heart-shaped at the base; *stem* branched; *flowers* numerous. — Hedges and waste ground ; common. A leafy plant, 2—3 feet high, with numerous small yellow flowers ; the lower leaves often have several small lobes running along the opposite sides of the stalks.— Fl. July, August. Annual.

* *L. pusilla* is a rare species, occasionally found in

cultivated ground. The *stems* in this species are leafless, 6—8 inches high, swelling and hollow upwards, and terminating in small yellow *flowers.*

CICHORIUM INTYBUS (*Wild Chicory, or Succory*).

13. CICHÓRIUM (*Chicory*).

1. *C. Íntybus* (Wild Chicory, or Succory).— Well distinguished by its tough, twig-like *stems,* along which are ranged large, blue, sessile *flowers.*—Sir Jas. E. Smith, the distinguished founder of the Linnean Society, seldom saw the delicate blue flowers of the Wild Succory without recalling to mind, that, when in infancy their beauty caught his eye, and attracted his admiration above most others, he tried in vain to pluck them from the stalk. In an introductory lecture at the Royal

Institution, he thus alludes to this circumstance :—
" From the earliest period of my recollection, when
I can just remember tugging ineffectually with all my
infant strength at the tough stalks of the wild Succory,

ARCTIUM LAPPA (*Common Bur-dock*).

on the chalky hills about Norwich, I have found the study of nature an increasing source of unalloyed pleasure, and a consolation and a refuge under every pain."

II. CYNAROCÉPHALÆ.—*Thistle Group*.

14. ARCTIUM (*Bur-dock*).

1. *A. Lappa* (Common Bur-dock).—The only British species.—Waste places; common. A large and stout herbaceous plant, remarkable for the picturesque character of its wavy *leaves*, (which are often introduced by artists into the frontispieces of their landscapes,) and for the round heads of purple *flowers*, enclosed in a globular *involucre* of hooked scales, by help of which the seeds are attached to the coats of animals, and conveyed to a distance. The scales are often interwoven with a white cottony substance.—Fl. July, August. Biennial.

15. SERRÁTULA (*Saw-wort*).

S. tinctória (Common Saw-wort).—The only British species.—Pastures; frequent. A slender plant, 1—2 feet high, with a stiff erect *stem*, slightly branched above, deeply cut and serrated *leaves*, and small terminal *heads* of purple *flowers;* the outer scales of the *involucre* are smooth, and close pressed, the inner tinged with purple. —Fl. August. Perennial.

16. SAUSSÚREA.

1. *S. alpína* (Alpine Saussurea).—The only British species, found on Snowdon and the Highland mountains. —The *stem* is from 8—12 inches high ; the *leaves* cottony beneath ; and the *flowers* light purple, in a terminal cluster.—Fl. August. Perennial.

SERRATULA TINCTORIA (*Common Saw-wort*).

CARDUUS NUTANS (*Musk Thistle*).

17. CARDUUS.

1. *C. nutans* (Musk Thistle).—*Heads* solitary, drooping ; *scales* of the *involucre* tapering to a rigid point, cottony, the outer ones bent back ; *stem* winged by the thorny leaves.—Waste places ; common.   A very hand-

some plant, about 2 feet high, with a furrowed cottony stem, deeply lobed thorny leaves, which are downy on the veins beneath, and large deep purple flowers, to which the radiated involucre is a very ornamental appendage. This is sometimes called the Scotch Thistle, but incorrectly. The upper part of the flower-stalk is nearly bare of leaves, and the flower itself has a powerful odour.—Fl. June—August. Biennial.

2. *C. acanthoídes* (Welted Thistle).—*Heads* clustered, round ; *scales* of the *involucre* linear, thorny, spreading, or erect ; *stem* winged by the thorny leaves.—Waste places ; common. A branched, very thorny plant, 3—4 feet high, with small heads, of deep purple, or sometimes white *flowers*.—Fl. June, July. Annual.

3. *C. tenuiflórus·* (Slender-flowered Thistle.)—*Heads* clustered, cylindrical ; *scales* of the *involucre* thorny, erect ; *stem* winged by the thorny leaves, which are cottony beneath.—Waste places, especially near the sea ; very common in the West of England. Well distinguished by the small heads of pink *flowers*, and the very long erect *scales* of the involucre. The stems are 2—4 feet high, and bear all the flowers at the summit.—Fl. June, July. Annual.

* *C. Mariánus* (Milk Thistle), is a stouter plant than either of the preceding, and is distinguished at once by the white veins on its *leaves*, from which it derives its name. It grows in waste places, but is not common.—Fl. June, July. Biennial.

18. Cnicus (*Plume-Thistle*).

1. *C. lanceolatus* (Spear Plume-Thistle).—*Heads* mostly solitary, stalked, egg-shaped ; *scales* of the involucre, thorny, spreading, woolly ; *stem* winged by the thorny leaves, the lobes of which are 2-cleft.—Waste places and hedges ; common. This is more like the Cotton-Thistle (*Onopordum*) than any other species of this genus. It grows 3—4 feet high ; the *leaves* are downy beneath,

and the *heads* of *flowers*, though not so large as those of the *Cotton-Thistle*, have the same dull purple hue.— Fl. July—September. Biennial.

CNICUS PALUSTRIS (*Marsh Plume-Thistle*).

2. *C. palustris* (Marsh Plume-Thistle).—*Heads* clustered, egg-shaped ; *scales* of the involucre closely pressed, pointed ; *stem* winged by the thorny leaves.—Moist meadows and borders of fields ; very common. The tallest of the British Thistles, 4—10 feet high, consisting of a single, stout, hollow *stem*, which is branched towards the summit, and bears numerous clusters of rather small, deep purple (sometimes white) *flowers*. The *leaves* are thickly armed with short thorns, which

are often of a brownish hue.—Fl. July, August. Biennial.

3. *C. arvensis* (Creeping Plume-Thistle).—*Heads* of flowers numerous, stalked, egg-shaped ; *scales* of the involucre closely pressed, pointed, but scarcely thorny ; *stem* not winged ; *root* creeping.—Border of fields ; very common. A handsome weed, about 2 feet high ; the flowers, which grow in a corymbose manner, are of a light purple colour, and smell like those of the *Musk Thistle.*—Fl. July. Perennial.

4. *C. pratensis* (Meadow Plume-Thistle).—*Heads* of flowers mostly solitary ; *stem-leaves* few, soft, and wavy. —Moist meadows ; not general. A small plant, 12— 18 inches high, with a cottony stem, bearing a few leaves, and rarely more than one small purple flower.—Fl. July. Perennial.

5. *C. acaulis* (Dwarf Plume-Thistle). — *Heads* of flowers solitary and stemless, or nearly so.—Dry gravelly or chalky pastures ; not general, but in some places very abundant, and a pernicious weed. A low plant, consisting of a few thorny *leaves*, and a single, stemless, purple *flower*, by which character it is readily distinguished from all the rest of the Thistle Tribe.—Fl. July, August. Perennial.

* Less common species of *Cnicus* are *C. eriöphorus* (Woolly-headed Plume-Thistle), distinguished by the thick white wool which clothes the *scales* of the very large *flowers : C. tuberosus* (Tuberous Plume-Thistle), which grows only in Wiltshire, an erect single-stemmed plant, with a single, large purple *flower : C. heterophyllus* (Melancholy Plume-Thistle), a mountain plant, with an erect, cottony *stem*, and a single, handsome, purple *flower.*

19. ONOPORDUM (*Cotton-Thistle*).

1. *C. Acanthium* (Scotch Thistle).—The commonest of the thistle-tribe, abounding in waste ground, and

by road sides, in most soils.   The *involucre* is globose,
with the *scales* spreading in all directions ; the *stem* is
winged with the rough cottony *leaves ;* the *flowers* are
large, of a dull purple hue, and mostly solitary, or but
slightly clustered, at the ends of the branches.   The
seeds are the favourite food of goldfinches, which may
often be seen in autumn fluttering like humming-birds
round the withered stalks.   This species is the true
Scotch Thistle.—Fl. July, August.   Biennial.

ONOPORDUM ACANTHIUM (*Scotch Thistle*).

CARLINA VULGARIS (*Common Carline Thistle*).

20. CARLÍNA (*Carline Thistle*).

1. *C. vulgáris* (Common Carline Thistle).—The only British species, growing on dry heaths, especially near the sea.—Readily distinguished from every other British Thistle by the long *inner scales* of the *involucre*, which are straw-coloured and glossy, and spread in a radiate manner so as to resemble petals. In dry weather they lie flat, but when the atmosphere is moist, they rise and form, as it were, a pent-house over the plants. Their

texture is like that of the flowers called "Everlasting;" hence, they scarcely alter their appearance when dead, and as the whole plant is remarkably durable, they often retain their form and position till the succeeding spring. In Germany, France, and Italy, the large white flower of one species, *C. acaulis*, is often nailed upon cottage doors, by way of a hygrometer, as it closes before rain.—Fl. June. Biennial.

21. CENTAURÉA (*Knap-weed, Blue-bottle*).

1. *C. nigra* (Black Knap-weed).—Outer scales of the *involucre* egg-shaped, fringed with spreading bristles; *lower leaves* toothed, often with a few small lobes at the base, *upper* narrow, tapering; *flowers* with or without a ray; *pappus* very short, tufted.—Meadows and corn-fields; common. A tough-stemmed plant, 1—2 feet high, with heads of dull purple *flowers*, which are remarkable for the brown, or almost black, hue of the scales of the *involucre*. This plant is popularly known by the name of *Hard-head*.—Fl. June—August. Perennial.

2. *C. Scabiósa* (Greater Knap-weed).—Outer *scales* of the *involucre* egg-shaped, somewhat downy, fringed; *leaves* pinnatifid, roughish, segments tapering to a point. — Meadows and corn-fields; common. Larger and stouter than the last, from which it is distinguished by the brighter hue of its handsome radiate flowers, and the light-coloured fringe on the scales of the involucre. —Fl. July, August. Perennial.

3. *C. Cýanus* (Corn Blue-bottle).—Outer *scales* of the *involucre* deeply toothed; *leaves* very narrow, slightly toothed, cottony.—Corn-fields; frequent. One of the prettiest of flowers, and well meriting the distinctive name often given to it, of *Corn-flower*. The flowers are bright blue, with dark anthers, and when mixed with Poppies and yellow Ox-daisies, form as brilliantly-coloured a bouquet as can be imagined. Children often

string the outer florets by help of a needle and thread, and after bringing the ends together, press them in a book, to form a wreath, which retains its colour for a long while. Recently expanded flowers should be chosen, or their colour will fade. The juice of the flowers, expressed and mixed with cold alum-water, may be used in water-colour drawing. Rose-coloured, white, and dark purple varieties are commonly to be met with in gardens, and the last are occasionally found in a wild state.—Fl. July, August; and in turnip-fields again in October and November. Annual, or Biennial.

c. CYANUS *and* c. CALCITRAPA (*Corn Blue-bottle and Star Thistle*).

A A

4. *C. Calcítrapa* (Common Star-Thistle).—*Scales* of the *involucre* ending in a long and stiff thorn.—Gravelly and sandy places in the South of England. Well marked by its purplish *flowers*, which are armed below with spreading thorns, and resemble in figure the cruel iron instrument, named a *Caltrops*, which is used in war to lame horses, being thrown on the ground where it is expected that cavalry will pass. The instrument is so constructed, that, in whatever position it lies, one point sticks upwards.—Fl. July, August. Annual.

* *C. Jácea* (Brown Radiant Knap-weed), is a very rare species, with purple *flowers*, the outer scales of the *involucre* being pinnatifid, the inner irregularly jagged.

III. TUBIFLÓRÆ.—*Tansy Group.*

22. BÍDENS (*Bur-Marigold*).

1. *B. cernua* (Nodding Bur-Marigold).—*Heads* of flowers drooping; *leaves* serrated, undivided; bristles of the fruit 3—4.—Watery places; frequent. A somewhat succulent plant, 1—2 feet high, with narrow, serrated, smooth leaves, and button-like, drooping heads of brownish yellow flowers, at the base of which are several leafy bracts. The fruit is oblong, and terminates in several stiff bristles, each of which is thickly set with minute points, which are turned back like the barbs of an arrow, so as to take a firm hold on the coat of any animal which comes in contact with them.—Fl. July—September. Annual.

2. *B. tripartíta* (Trifid Bur-Marigold). — *Heads* of flowers nearly erect; *leaves* 3-parted.—Watery places; common. Distinguished from the last by its somewhat

smaller *heads* of flowers, 3-parted *leaves,* and by having 2—3 bristles on the fruit.—Fl. July—September. Annual.

BIDENS TRIPARTITA (*Trifid Bur-Marigold*).

23. EUPATORIUM (*Hemp-Agrimony*).

1. *E. cannábinum* (Common Hemp-Agrimony). — The only British species.—Moist shady places ; common. A tall downy plant, 3—6 feet high, with a reddish *stem,* 3—5 parted *leaves,* and terminal corymbs

of small crowded *heads* of dull lilac *flowers*, which are remarkable for their very long, deeply cloven *styles.*—Fl. July, August. Perennial.

EUPATORIUM CANNABINUM (*Common Hemp-Agrimony*).

24. Chrysócoma (*Goldylocks*).

1. *C. Linosýris* (Flax-leaved Goldylocks).—The only British species.—Limestone cliffs ; very rare. A herbaceous plant, 12—18 inches high, with erect, simple *stems*, which are thickly set with smooth, linear *leaves*, and bear a few heads of yellow *flowers* at the extremity. —Fl. August, September. Perennial.

25. Diótis (*Cotton-weed*).

1. *D. marítima* (Sea-side Cotton-weed).—The only species.—Sandy sea-shores ; rare. The *roots* run deeply

into the sand ; the *stems*, which are about a foot high, are thickly set with oblong, blunt *leaves*, which, as well as the rest of the plant, are covered with thick white cotton, and almost hide the small terminal *heads* of yellow *flowers.*—Fl. August, September. Perennial.

TANACETUM VULGARE (*Common Tansy*).

26. TANACÉTUM (*Tansy.*)

1. *T. vulgáré* (Common Tansy).—The only British species.—Hedges, and waste ground ; common. Growing 2—3 feet high, and well distinguished by its deeply twice pinnatifid, cut *leaves*, and terminal corymbs of bright yellow, button-like *flowers*. The whole plant is

bitter, and aromatic, and is not only used in medicine, but forms the principal ingredient in the nauseous dish called *Tansy pudding.*—Fl. August.   Perennial.

ARTEMISIA ABSINTHIUM (*Common Wormwood*).

27. ARTEMISIA (*Wormwood, Mugwort*).

1. *A. Absinthium* (Common Wormwood). — *Leaves* with bluntish segments, twice pinnatifid, silky on both

sides; *heads* hemispherical, drooping.—Waste ground; common. A bushy plant, with silky stems and leaves, and panicles of numerous small heads of dull yellow flowers. The whole plant is bitter and aromatic, and is much used in the rural districts, where it abounds, as a tonic.—Fl. July—September. Perennial.

2. *A. vulgaris* (Mugwort).—*Leaves* pinnatifid, with acute segments; white with down beneath; *heads* oblong.—Hedges, and waste places; common. Taller, and more slender than the last; well distinguished by the *leaves* being green above and white below, and by the absence of aromatic odour.—Fl. July—September. Perennial.

3. *A. maritima* (Sea Wormwood). — *Leaves* twice pinnatifid, downy on both sides; *heads* in racemes, oblong.—Salt marshes; frequent. Somewhat resembling *A. Absinthium*, but smaller and well distinguished by the above characters. The clusters of flowers are sometimes drooping, sometimes erect.—Fl. July—September. Perennial.

* *A. campestris* (Field-Wormwood) is a rare species, growing on sandy heaths in Norfolk and Suffolk. In this species the segments of the *leaves* terminate in points, and the *stems*, until flowering, are prostrate.

28. GNAPHALIUM (*Cud-weed*).

1. *G. dioicum* (Mountain Cud-weed).— *Stamens* and *pistils* on separate plants.—Mountain heaths, frequent. A pretty little plant, 3—6 inches high, with oblong *leaves*, which are broadest towards the end, green above, cottony below; the *heads* of flowers grow 4—6 together, and are rendered conspicuous by the white or rose-coloured *involucre*, which is of the texture commonly termed *everlasting*.—Fl. July, August. Perennial.

* The *White Everlasting* of gardens is *G. margaritaceum*, a much larger plant. It is occasionally found in

GNAPHALIUM DIOICUM (*Mountain Cud-weed*).

situations where it is apparently wild, but is not considered to be indigenous.

2. *G. uliginosum* (Marsh Cud-weed).—*Stems* much branched, woolly ; *leaves* very narrow, downy, overtopping the clustered, terminal heads.—Wet sandy places, especially where water has stood during winter, common. A small plant, 3—6 inches high, rendered conspicuous by its tufted white stems and leaves, and by the glossy, yellowish-brown scales of its small clustered flowers.—Fl. August, September. Annual.

* Two other species of Cud-weed are found in Scotland : *G. sylváticum* (Highland Cud-weed), a cottony plant, with a simple *stem*, bearing its heads of *flowers* in a leafy *spike ;* this species is also found in gravelly pastures in England : and *G. supinum* (Dwarf Cud-

weed), common on the summits of the Highland mountains, also a cottony plant, but rarely exceeding 2—3
inches in height, with tufted *leaves,* and flowering *stems*
almost bare of leaves.

FILAGO GERMANICA (*Common Filago*).

29. FILÁGO.

1. *F. Germánica* (Common Filágo).—*Stem* cottony,
erect, terminating in a globular assemblage of *heads,*
from the base of which rise two or more *flower-stalks,*
which are prolific in like manner.—Dry gravelly places,
common. A singular little plant, 6—8 inches high,

well distinguished by the above character. From this curious mode of growth, the plant was called by the old botanists *Herba impia* (the undutiful herb), as if the young shoots were guilty of disrespect by overtopping the parent.—Fl. June, July. Annual.

2. *F. mínima* (Least Filágo).—*Stem* erect, repeatedly forked ; *leaves* very narrow, cottony, pressed to the stem ; *heads* conical, in lateral and terminal clusters, shorter than the leaves.—Dry gravelly places, common. Yet smaller than the last, growing 4—6 inches high, with cottony stems and leaves, and brownish-yellow flowers.—Fl. July, August. Annual.

* A third species of *Filágo* (*F. Gállica*) occurs in several parts of England and Scotland, and differs from the last in bearing *leaves* much longer than the *flowers*.

30. PETASÍTES (*Butter-Bur*).

1. *P. vulgáris* (Common Butter-Bur).—The only British species. — Marshy meadows, not unfrequent. The largest, and, when it abounds, the most pernicious, of all the weeds which this country produces. The *flowers*, which are of a dull lilac colour, appear early in spring, and are succeeded by downy, kidney-shaped *leaves*, 1—3 inches in diameter, which by shading the ground check the growth of all other plants. "The early blossoming of this rank weed induces the Swedish farmers to plant it near their bee-hives. Thus we see in our gardens the bees assembled on its affinities, *P. alba* and *P. fragrans*, at a season when scarcely any other flowers are expanded."—*Sir W. J. Hooker.* These two last species are common in shrubberies, almost hiding the ground with their broad leaves, thriving beneath the shade of trees and shrubs, but overpowering all herbaceous plants, and eventually, it is said, even the shrubs themselves.—Fl. April, May. Perennial.

PETASITES VULGARIS (*Common Butter-Bur*).

TUSSILAGO FARFARA (*Colt's-foot*).

IV.—RADIATÆ.—*Daisy Group.*

31. TUSSILÁGO (*Colt's-foot*).

1. *T. Fárfara* (Colt's-foot).—The only British species, abounding in clayey fields, where it is a very pernicious weed. The *flower-stalks*, which spring directly from the roots, are covered with scale-like *bracts*, and bear each a single yellow *flower*, with numerous narrow rays ; the

leaves, which do not appear until the flowers have withered, are roundish heart-shaped and angular, with dark teeth, and are covered with cottony down. The *heads* of *flowers* droop before expansion, and the stalks after flowering lengthen considerably. The cotton of the leaves was formerly used as tinder, and the leaves themselves afford a rustic remedy for coughs.—Fl. March, April. Perennial.

32. Erígeron (*Flea-bane*).

1. *E. acris* (Blue Flea-bane).—*Branches* erect, rough, alternate, bearing single heads; *leaves* narrow, entire, blunt.—Dry places and walls, not common. A much branched plant, 6—18 inches high, with small heads of inconspicuous flowers, of which the inner florets are yellowish, the outer dull blue. The pappus is very long and tawny.—Fl. August. Biennial.

* *E. alpínus* (Alpine Flea-bane) occurs on the Highland mountains, growing 3—5 inches high, each stem bearing a single *flower*, the outer *florets* of which are light purple : *E. Canadensis* (Canada Flea-bane) grows as a weed in waste ground and on old walls about Chelsea and elsewhere; it has somewhat of the habit of Groundsel, and bears small *heads* of dingy yellow *flowers.*

33. Aster (*Starwort*).

1. *A. Tripólium* (Sea Starwort).—The only British species, abundant in salt marshes and on sea-cliffs. A stout succulent plant, 2—3 feet high, with long, smooth, fleshy *leaves*, and corymbs of large handsome *heads* of *flowers*, the inner *florets* of which are yellow, the outer purple. In salt marshes the whole plant is often covered with mud, which gives it an unsightly appearance, but

when growing on sea-cliffs, it is a highly ornamental plant.—Fl. August, September.   Perennial.

ASTER TRIPOLIUM (*Sea Starwort*).

34. Solidágo (*Golden-rod*).

S. Virgaurea (Golden-rod).—The only British species, common in dry woods.—An erect, scarcely branched plant, 2—3 feet high, with roughish, angular *stems*, simple, serrated *leaves*, which gradually become narrower the higher they are on the stem ; and conspicuous, terminal, crowded clusters of small bright yellow *flowers*.

On mountainous heaths a variety occurs with very short *stems* and large *leaves* and *flowers*.—Fl. July—September. Perennial.

SOLIDAGO VIRGAUREA (*Golden-rod*).

35. Senécio (*Groundsel, Ragwort*).

1. *S. vulgáris* (Common Groundsel).—*Flowers* without rays, in crowded clusters ; *leaves* half-embracing the stem, deeply lobed and toothed.—A common weed in

cultivated ground ; a favourite food of many small birds.—Fl. all the year round.   Annual.

SENECIO VULGARIS (*Common Groundsel*).

2. *S. sylváticus* (Mountain Groundsel).—*Flowers* with a few rays, which are rolled back and inconspicuous, or often wanting ; *leaves* pinnatifid, with narrow lobes, toothed.—Gravelly places, common. Distinguished from the last by its larger size, its more copiously cut, often hoary *leaves*, and its conical rather than cylindrical heads of dull yellow flowers. The *stem* is branched, 1—2 feet high.—Fl. July—September.   Annual.

3. *S. Jacobœa* (Common Ragwort).—*Flowers* with spreading rays; leaves pinnatifid, with smaller lobes at the base.—Meadows and wet places, common.—*Stem* erect, 2—3 feet high, *flowers* large, bright yellow, corymbose.—Fl. July—September. Perennial.

SENECIO JACOBÆA (*Common Ragwort*).

4. *S. aquáticus* (Marsh Ragwort).—*Flowers* with spreading rays; *lower leaves* undivided, toothed, *upper* with a few oblong lobes near the base.—Wet places, common. Resembling the last, but readily distinguished by its less divided leaves. — Fl. July — September. Perennial.

5. *S. tenuifolius* (Hoary Ragwort).—*Flowers* with spreading rays; *leaves* pinnatifid, with very narrow segments, downy beneath.—Dry banks in a limestone

or chalky soil, not common. Of about the same size as
the last, but distinguished by its sending up numerous
cottony stems from the same root, and by its regularly
divided leaves, the segments of which are slightly rolled
back at the edges.—Fl. July, August. Perennial.

Less common species of *Senécio* are *S. viscósus* (Viscid
Groundsel), which approaches *S. sylváticus* in habit,
and is distinguished by its being clothed with viscid
hairs : *S. squálidus* (Inelegant Ragwort), growing about
a foot high, with large bright yellow flowers, abundant
on old walls about Oxford, and Bideford, Devon; whence
it derived its name is not clear, inasmuch as it is by
far the prettiest British species : *S. paludósus* (Great
Fen Ragwort), a large aquatic plant, 5—6 feet high,
with undivided *leaves,* growing very sparingly in the
east of England : and *S. Saracénicus* (Broad-leaved Ragwort), also with undivided leaves, but much smaller
than the last ; it is not considered to be indigenous.

36. Cinerária (Flea-wort).

C. palustris (Marsh Flea-wort).—Shaggy ; *stem* much
branched, hollow.—Fens of Norfolk and Cambridge,
rare. A stout plant, 2—3 feet high, with numerous
toothed leaves, which are wavy at the edges ; flowers
bright yellow.—Fl. June, July. Perennial.

C. campestris (Field Flea-wort). — Shaggy ; *stem*
simple ; *root-leaves* oblong, nearly entire, *stem-leaves*
narrow, tapering.—Chalky downs in the East of England, not common. A small plant, 6—8 inches high,
with a few small stem leaves, and umbels of yellow
flowers.—Fl. May, June.

37. Dorónicum (*Leopard's-bane*).

1. *D. Pardalianches* (Great Leopard's-bane).—*Lower
leaves* heart-shaped, toothed, on long stalks, *upper* with

two ears at the base, embracing the stem.—Damp hilly woods, rare, a doubtful native. Stem 2—3 feet high, erect, solitary, hollow, hairy; leaves soft; heads of flowers yellow, the earlier ones overtopped by the later. —Fl. May—July. Perennial.

* *D. plantagineum* (Plantain-leaved Leopard's-bane) differs from the last in having egg-shaped *leaves* and solitary *heads* of *flowers*. It is very rare, and is not considered to be indigenous.

38. Ínula (*Elecampane, Ploughman's Spikenard*).

1. *I. Helénium* (Elecampane).—*Leaves* oblong or egg-shaped, wrinkled, downy beneath, toothed, upper ones embracing the stem; *scales* of the *involucre* egg-shaped, downy.—Moist pastures, not common. A stout plant, 3—5 feet high, with very large leaves and a few terminal very large heads of bright yellow flowers. The root contains a white starchy powder, named Inuline, a volatile oil, a soft acrid resin, and a bitter extract; it is used in diseases of the chest and lungs, and furnishes the *Vin d'Aulnée* of the French.—Fl. July, August. Perennial.

2. *I. Conýza* (Ploughman's Spikenard).—*Leaves* narrow, egg-shaped, downy, toothed; *heads* of flowers panicled; *scales* of the involucre rolled back.—Hedges, principally on a limestone or chalky soil, frequent. Distinguished by its dull green foliage, numerous heads of dingy yellow flowers, the rays of which are inconspicuous, and by the leaf-like scales of the involucre, which are rolled back.—Fl. July—September. Perennial.

3. *I. crithmoídes* (Golden Samphire).—*Leaves* very narrow, fleshy, smooth, blunt, or 3-pointed. — Salt marshes, and sea-cliffs, rare. Well distinguished from every other British plant by its fleshy leaves and large

yellow flowers, which grow singly at the extremity of
the branches.—Fl. July, August.   Perennial.

INULA HELENIUM (*Elecampane*).

39. PULICARIA (*Flea-bane*).

1. *P. dysentérica* (Common Flea-bane).—*Stem* woolly;
leaves oblong, heart or arrow-shaped at the base, embracing the stem ; *scales* of the *involucre* bristle-shaped.
—Watery places, common ; rare in Scotland.   From

1—2 feet high, growing in masses, and well marked by its soft hoary foliage and large flat heads of bright yellow *flowers*, those of the ray being very numerous, narrow, and longer than the disk.—Fl. August. Perennial.

PULICARIA DYSENTERICA (*Common Flea-bane*).

2. *P. vulgáris* (Small Flea-bane).—*Stem* hairy ; *leaves* narrow, tapering, hairy.—Sandy heaths, where water has stood, not common ; said to be not found in Scotland or Ireland. Resembling the last, but not above half the size, nor by any means so hoary.—Fl. September. Annual.

BELLIS PERENNIS (*Common Daisy*).

40. BELLIS (*Daisy*).

1. *B. perennis* (Common Daisy).—The only British species, too well known and admired to need any description or comment.—Fl. nearly all the year round. Perennial.

41. CHRYSÁNTHEMUM (*Ox-eye*).

1. *C. Leucánthemum* (White Ox-eye).—*Florets* of the ray white; *lower leaves* stalked, *upper* sessile, pinnatifid at the base.—Meadows, abundant. Almost as well known as the common daisy. A great favourite of children, who string the flowers on a stout grass-straw, or bit of wire, and make a very fair imitation of a soldier's feather. It is said to be destructive to fleas.—Fl. June, July. Perennial.

CHRYSANTHEMUM LEUCANTHEMUM (*White Ox-eye*).

2. *C. ségetum* (Yellow Ox-eye, Corn Marigold). —
Florets of the ray yellow ; *leaves* clasping the stem, oblong, acute, toothed, glaucous.—Corn-fields, abundant,
but local. Whole plant remarkably smooth and glaucous ; the *flowers* are large, of a brilliant yellow, and
contrast beautifully with Poppies and Blue-bottles.—
Fl. June, July, and, in summer-ploughed fields, again in
October and November. Annual.

42. PÝRETHRUM (*Feverfew*).

1. *P. Parthénium* (Common Feverfew). — *Leaves*
stalked, pinnate ; *leaflets* pinnatifid, and deeply cut ;

stem erect ; *flowers* corymbose. — Hedges **and waste**
ground, common.   Well marked by its repeatedly cut,
curled, delicate green leaves, and its numerous small
heads of flowers, of which the central ones are yellow,
the outer white.   The leaves are conspicuous in mid-
winter, and the whole plant has a powerful and not
unpleasant odour, which is said to be particularly offen-
sive to bees.   The English name is a corruption of
Febrifuge, from its tonic properties.—Fl. July, August.
Perennial.

PYRETHRUM INODORUM (*Corn Feverfew, Scentless May-weed*).

2. *P. inodórum* (Corn Feverfew, Scentless May-weed).
—*Leaves* sessile, repeatedly cut into numerous hair-like
segments ; *stem* branched, spreading ; *flowers* solitary.

—Corn-fields, common. Of a very different habit from the last, but resembling it in the colour of the flowers, which are, however, much larger, and are remarkable for their very convex disk. When growing near the sea, its leaves become fleshy, when it is called *P. maritimum* (Sea Feverfew).—Fl. July—October. Annual.

43. MATRICARIA (*Wild Chamomile.*)

1. *M. Chamomilla* (Wild Chamomile).—The only British species, growing in corn-fields, not uncommon. The *heads* of flowers are white, with a yellow disk. It may be distinguished from *Pýrethrum inodórum* and *Ánthemis Cótula* by the scales of the *involucre* being not chaffy at the margin, and by the *receptacle* of the florets being hollow.—Fl. June—August. Annual.

44. ÁNTHEMIS (*Chamomile*).

1. *A. nóbilis* (Common Chamomile).—*Stems* prostrate; *leaves* repeatedly cut into hair-like segments, slightly downy.—Heaths, abundant. Well distinguished by its solitary *heads* of flowers, which droop before expansion, and by its pleasant aromatic smell, which resembles that of fresh apples, whence it derived its name of *Chamomile*, signifying in Greek, *ground apple*. The whole plant is very bitter, and is valuable in medicine for its tonic properties.—Fl. August. Perennial.

2. *A. Cótula* (Stinking Chamomile). — *Stem* erect, branched; *leaves* repeatedly cut into hair-like segments, smooth.—Waste places, common. Distinguished from the last by its strong disagreeable odour, and upright stems. The heads of flowers are solitary, coloured as in the last, but larger. The juice is very acrid, and is said to blister the hands of those who gather it.—Fl. July, August. Annual.

* Less common species of Chamomile are *A. marítima* (Sea Chamomile), which has repeatedly-cut fleshy *leaves*,

which are somewhat hairy ; on the sea-coast, very rare : *A. arvensis* (Corn Chamomile), the deeply cut *leaves* of which are white with down ; these two have *white flowers* with a *yellow disk :* and *A. tinctoria* (Ox-eye Chamomile), which has downy, much divided leaves, and large bright yellow *flowers,* resembling *Chrysánthemum ségetum.*

ANTHEMIS NOBILIS (*Common Chamomile*).

ACHILLEA PTARMICA (*Sneeze-wort*).

45. ACHILLÉA (*Yarrow*).

1. *A. Millefólium* (Common Yarrow, Milfoil). — *Leaves* twice pinnatifid, woolly, or slightly hairy ; *leaflets* cut into hair-like segments ; *flowers* in dense terminal corymbs.—Waste ground, frequent. A common road-side plant, with very tough, angular stems, 1—2 feet high, and corymbs of small, white, pink, or purplish flowers, which, by an unpractised eye, might be supposed to belong to an umbelliferous plant. It has a strong and slightly aromatic odour, and is said to have the property of healing wounds.—Fl. June—September. Perennial.

2. *A. Ptármica* (Sneeze-wort). — *Leaves* undivided, very narrow, and tapering to a sharp point, serrated.— Meadows, and waste ground, not uncommon. Somewhat taller and slenderer than the last, from which it may be at once distinguished by its undivided leaves and larger heads of *flowers*, of which both the *disk* and *ray* are white. The pounded leaves have been used as snuff; hence its name.—Fl. July, August. Perennial.

* *A. tomentósa* (Woolly Yellow Yarrow) is very rare, and a doubtful native; its *leaves* are repeatedly divided, and woolly, the flowers bright yellow.

END OF VOL. I.

R. CLAY, PRINTER, BREAD STREET HILL.

Printed in the United States
By Bookmasters